salt

ARTS AND TRADITIONS OF THE TABLE: PERSPECTIVES ON CULINARY HISTORY
ALBERT SONNENFELD, SERIES EDITOR

ARTS AND TRADITIONS OF THE TABLE:
PERSPECTIVES ON CULINARY HISTORY

ALBERT SONNENFELD, SERIES EDITOR

PIERRE LASZLO

Translated by Mary Beth Mader

salt grain of life

columbia university press

new york

Columbia University Press wishes to express its appreciation for assistance given by the government of France through the Ministère de la Culture in the preparation of this translation.

The quotation in the foreword from Pablo Neruda's "Salt" comes from *Selected Odes of Pablo Neruda*, edited and translated by Margaret Sayers Peden. Copyright © Regents of the University of California, © Fundacion Pablo Neruda.

The author's Web site can be visited at www.pierrelaszlo.net.

Columbia University Press
Publishers Since 1893
New York   Chichester, West Sussex

Translation copyright © 2001 Columbia University Press
Copyright © 1998 Hachette Littératures

Library of Congress Cataloging-in-Publication Data
Laszlo, Pierre.
[Chemins et savoirs du sel. English]
Salt : grain of life / by Pierre Laszlo ;
translated by Mary Beth Mader.
p. cm.
Includes bibliographical references.
ISBN 0–231–12198–9
1. Salt—History. I. Title.
TN900 .L3713   2001
553.6'3—dc21   2001–028227

Casebound editions of Columbia University Press books are
printed on permanent and durable acid-free paper.

Printed in the United States of America
c 10 9 8 7 6 5 4 3 2 1

TO MY SALTY AND SPICY WIFE

(eew!)

contents

foreword

Salt
I saw the salt
In this shaker
In the salt flat
I know
You will never believe me
but it sings
the salt sings . . .
                    —Pablo Neruda

Pierre Laszlo's delightfully informative *Salt* could not have appeared at a more opportune time.

Consumers' hypertension notwithstanding, salt has reassumed in the cuisine and prosperous economies of a new millennium the central role it had occupied for centuries. Salt has even become a staple ingredient of desserts, so that the sweet trolley now blends the two savors: a brilliant chocolate tart by a great Parisian *pâtissier* contains enough *fleur de sel* in the pastry crust to enhance the blackest chocolateness while tempering its acidity. And salt contributes to the sought-after texture of puff pastry. So too a black-pepper ice cream benefits from saline arousal!

None of these salt-enhanced gourmet delights comes cheaply, what with a culinary repertory including sea salt from the Andes, pink Peruvian sea salt, smoked Danish salt, and Maldon sea salt from Essex, not to mention the better-known *sel gris* and *fleur de sel*. These now retail for up to $85.00 a pound! At the cash register, it never rains, it pours!

A much-acclaimed chemist and a master pedagogue, Pierre Laszlo shows scientifically the indispensability and effects of salt in the cooking process. But, a true polymath, Laszlo is also an etymologist, economist, historian, and pop philosopher. The reader will, as I was, be amazed at the epicentrality of salt in history, economics, and philology, convincingly demonstrated here in an eminently readable fashion.

The state monopoly salt tax of yore has become a new value-added tax as designer salts are sprinkled on gourmet dishes in upscale surroundings. No trip to Salzburg, Salt Lake City, and Salisbury will ever be the same after the illumination this book provides.

—*Albert Sonnenfeld, series editor, Arts and Traditions of the Table: Perspectives on Culinary History*

preface

As I learned of the preparation of an English-language edition of this book, I took a trip from Ithaca to San Francisco on the occasion of the spring break at Cornell, where I was serving as a visiting professor. As my plane took off from the Tompkins County airport, I could see extremely well the salt mine at Lansing, right on the shore of Cayuga Lake. The lake was sky blue, and a white plume of water vapor was coming off the chimney stack, rising vertically into the still air. Later in the day, after a transfer in Pittsburgh and just before landing at the San Francisco airport, I could see the salt flats at Newark, a little south of the airport, on the Bay shore:

> A long brown bar at the dip of the sky
> Puts an arm of sand in the span of salt.
> <div align="right">Carl Sandburg,
"Sketch," <i>Chicago Poems</i> (1916)</div>

Both salt plants, in upstate New York and in California, are now owned by the same corporation, Cargill. My trip across the continent had been framed by salt production, just as the westward expansion of the United States in the nineteenth century had been fueled by salt and made perilous by the occasional lack or scarcity of this precious, indispensable resource.

To quote Sandburg again and to amalgamate the migration West and the Civil War, the history of salt making in the United States, both craft and industry, is inseparable from "the spawning tides of men and women / Swarming always in a drift of millions to the dust of toil, the salt of tears / And blood drops of undiminishing war" ("Momus," *Chicago Poems* [1916]).

"Spawning tides of men": production of salt by evaporation from the shallow waters of San Francisco Bay was started as a direct consequence of the Gold Rush to provide the hordes invading California with their daily salt as well as a necessary meat preserver. After the 1850s, the saltworks in Newark, California, were owned first by the Arden Salt Company, predecessor of the Leslie Salt Company, which Cargill Salt purchased in 1978.

"And blood drops of undiminishing war": production of salt in upstate New York was one of the issues disputed during the Indian Wars at the turn of the nineteenth century. The salt springs near Syracuse eventually passed from control of the French to that of the colonists and then quickly to the state of New York, which to a large extent was rushed into existence by Congress in order to supervise and tax salt extraction, as I document in this book.

Thus the North American continent is, as it were, bracketed by salt, from Lansing, New York, to Newark, California. The geography of the United States is sprinkled with salt, and its history is steeped and pickled in salt, from the Welsh settlers in the mountains of North Carolina, desperate for salt, who carefully retrieved the precious substance from the floors of the little wooden cabins where they smoked ham; to the pioneers in Ohio and Michigan, where salt exploitation provided supplies for the wagons heading West along trails such as the Santa Fe and the Oregon; to the Mormons fleeing persecution in the East and in Illinois, who finally found shelter in Utah, on the shore of the Great Salt Lake. There, as in a number of locations on the North American continent, the salt discovered was a dousing rather than just a sprinkling.

The United States has entertained an original relationship with salt. Even though it fills a universal need, its cultural and technical status in the U.S. has been at variance, sometimes in a subtle way, sometimes more starkly, with its standing elsewhere, in Europe in particular. A people having made the Boston Tea Party their symbol of revolt against tyranny and of the struggle for independence would not tolerate too heavy a taxation of essentials such as salt, in older days, or, today, gasoline. To the American people, taxation remains an odious burden that politicians continue to use quite effectively, even at the dawn of the twenty-first century. The following pages chronicle the long-standing association between political power and monopolistic control of salt distribution.

Today, the world's largest corporation producing and distributing salt, for purposes as diverse as deicing wintry roads and refilling salt shakers in homes and restaurants, is American. A U.S. corporation made its fortune from its proprietary knowledge of how to use salt as the raw material for numerous chemicals: Dow Chemical, which built a thriving business on the chloralkali chain from electrolysis of brine, is today the leading chemical company in the United States and among the four or five leading manufacturers of chemical commodities worldwide. This book will reveal the slow, gradual, and almost insidious passage of the power exerted by seemingly innocent, lily-white salt, which our body requires in daily amounts of several grams, from the despotic rulers in medieval Europe to the multinational corporations presently running our global village.

The availability of salt meant survival to the pioneers trudging west, a source, no doubt, of stories galore, but it is the literary description of their daily lives once they were settled that has always captured my imagination. Long before I first set foot in the United States as a young man, I was already inhabiting an America of the mind. My adult life has been spent, in a deep sense, connecting personal experiences with the fantasy world that, as a young boy, had been made so vivid to me by writers who evoked explorers and trappers, settlers and ruffians, from the Midwest to Alaska. I was nurtured on Jack London and Jules Verne. Those writers gave me a longing for the simple life, their descriptions of the blank slate of the virgin continent incised in the throbbing, living tissue of my imagination. Hence in my native France I was beckoned to an America that I still find irresistible in its utopian stark blandness: a small group of people living on the edge of wilderness, getting their supplies, such as salt or gunpowder, from a trading post or a general store, with a post office serving as their link to the rest of the world.

Such reduction to the essentials, much the way salting concentrates the flavor of meat, has remained to me the main attraction of the small-town America I continue to adore, a taste I share with my American wife. I find it driving along U.S. Route 1 on the Pacific shore in California or in Oregon; or on the magnificent and to some extent still sparsely populated Outer Banks of North Carolina; or along the coast of Maine; or, and this is one of my favorite drives at any time of the year, between Burlington, Vermont, and Montreal; or, to indulge in yet another visual reference, my tendency to cobble together bits of Americana to create a landscape of utter simplicity and searing beauty, when traveling at random through the recesses of the Chesapeake Bay.

The short story is very much an American art form. It also preserves and enhances flavor. I am very fond of the short format, of the miniature. Hence my taste in fiction for writers such as Edgar Allan Poe, William Saroyan, John O'Hara, or Eudora Welty. Hence my taste for the exquisite essays of E. B. White and for the painting masterpieces of Saul Steinberg, to mention other and related art forms, also minimalist, also etched in wit and lyricism, also wary of sweep and pathos.

This affection accounts for the format of this book. It is a collection of small vignettes, each devoted to one of the many aspects of salt, each endeavoring to isolate a distinct sparkle in the constellation of multiple meanings mankind, from ages past, has loaded on to the white, tasty powder. I hope that the Americanness of the favored means of expression will help to make my brief congenial to American readers. I have provided them with a half-dozen additional segments, written specifically for this edition.

acknowledgments

Valuable information was provided to me by Mr. Patrick Arabeye, Dijon; Mrs. Diane Berrier, potter, Saint-Forget; Mr. Pierre Chiesa, Paris; Ian dell'Antonio, Kitt Peak Observatory, Arizona; Mr. Jean-Loup Fontana, Nice; Professor Sture Forsén, University of Lund; Mrs. Lella Gandini, University of Massachusetts, Amherst; Professor Raquel Gonçalves, University of Lisbon; Miss Elizabeth Heinzelmann, Bern; Mrs. Chantal Humbert, Souliers; Professor Waclaw Kolodziejski, Warsaw; Mr. Jean-Bernard Lacroix, Nice; Miss Chloé Laszlo, Utrecht; Professor Lester K. Little, Northampton, Massachusetts; Mr. Jérôme Lovy, Paris; Loïc and Hong Mahé, Nantes; Ilona and Hugh Morison, Edinburgh; Mrs. Danielle Musset, Alpes de Lumière, Salagon par Mane; Dr. Liamó Dochartaigh, University of Limerick; Professor Jean-Paul Poirier, Institut de physique du globe, Paris; Countess Bertrand de Saint-Seine, Longecourt-en-Plaine; Professor Sugawara Tadashi, University of Tokyo; Dr. Hervé This-Benckhard, Paris; Benoît Van Reeth, Besançon; Dr. Samir Zard, Ecole polytechnique. May all of these people accept my deepest gratitude for their assistance.

I greatly benefited from the very warm hospitality of the Fondation des Treilles during the last two weeks of April 1997. This stay proved tremendously useful for the assembly of the book: may this benefactor be assured of my thanks.

Benoît Chantre played the role of editor to perfection. He has my sincere gratitude. I was very fortunate in having Jennifer Crewe at the helm at Columbia University Press; she is a superb human being and a great editor as well. Her assistant, Jennifer Barager, took an essential role in ensuring the safe completion of many of the attendant tasks, and I am very grateful to her as well. The managing editor, Anne McCoy, retained her friendliness and her cool under fire. She too has a wonderful assistant in Godwin Chu, who had to collate the proofs in no time: I thank them both for outstanding work under deadline duress.

I am indebted for the appearance of this translation to the Bureau du livre français in New York, with its young and highly talented staff members. They effectively served as my agent, and I thank them for unflagging support of this book.

I am enormously grateful to all my friends and colleagues, too numerous for individual mention, in the Department of Chemistry at Cornell University—where some segments of this book were written and where parts of the translation were reviewed—for the warmth of their hospitality. They make me feel at home in their great department.

I am deeply grateful to Professor Albert Sonnenfeld for having welcomed this book into the collection he directs at Columbia University Press. I offer special thanks to my translator, Mary Beth Mader: she had to cope with the French language in several registers, some of them a little arcane, and it was a joy to interact with her during the course of her work. The copy editor, Sarah St. Onge, gave felicitous and harmonious wording to the whole text, which I gratefully acknowledge.

introduction

Each time I have taught freshman general chemistry, I have experienced a small, mischievous pleasure in giving students not solutions but scientific bait for thought through questions like "Why put salt on snowy roads?" or "Why salt the water for boiling eggs?" Convinced as they were of the split between home and school, of the separation between everyday life and knowledge, this surprised them.

Their pessimism made me think. Our teaching, on all levels, suffers from specialization, from the partitioning of knowledge into established disciplines. I have thus written this book for the public at large but also as a pedagogical utopia, as the dream of a multidisciplinary form of teaching that combines literary analysis, history, anthropology, biology, economics, art history, physics, political science, chemistry, ethology, linguistics, and so on. This pedagogical project at first prompted me to call the book a "treatise." (There is no lack of other subjects that merit similar treatment—gasoline, water, spices, jams, the moon, wood, flowers, wheat, oranges, and gardens are several possibilities—and I may perhaps one day attempt to write another short treatise, whether of the present sort or of another.) But this attempt most certainly does not aim to be as exhaustive as a learned treatise. I present it, rather, as an *ignorant* treatise.

Room might be made in the language for such a neologism. Where the learned treatise is erudite and sententious, the ignorant treatise is fantastical and whimsical. A learned treatise is written for poor students; an ignorant treatise is written for all students, that is, for us all. A learned treatise—for example, any of Georges Bruhat's monumental and unsurpassable works in physics—is difficult. Reading it requires note taking, if only to set out the steps of an argument or to follow a calculation. A learned treatise follows a linear logic, step by step. An ignorant treatise indulges in exploring byways, in finding shortcuts, and in seeking out the harmony of language with the world. I salute Francis Ponge as one of the masters at writing ignorant treatises. I owe another debt: Primo Levi wrote a superb book, *The Periodic Table*, based on Mendeleyev's classification of the elements. For reasons he took with him to the grave, he refrained from writing anything about chlorine and sodium. I have boldly resolved to fill that gap, as far I am able.

We must devise a poetics of knowledge that is also an ethics, for knowledge is not parceled out in finished products ready for delivery; it is won by an effort renewed daily. To understand the world is above all to be aware of the questions it asks us if we care to be attentive. The most mundane utilitarian object—salt, for example—has many lessons to teach us. Such is the legacy of one of the twentieth century's great intellectuals, Marcel Duchamp, and of his ready-mades. (Dare I remind the reader that he also made wordplay of his name? He called himself "Marchand du sel," i.e., salt merchant.)

As I said, the ignorant treatise proposes nomadism as its guiding principle. It must glean random scraps of the most diverse types of knowledge before gathering them into a tale to tell, for we are victims of dispersion, of explosion into little districts of hyperspecialization. The ignorant treatise begins by collecting these scraps, this dust, a little like powdered salt, which, in the light, under a magnifier, would display small flashes of lightning or sparks: salt bursts, the crystalline burst of salt, bursting actions by and for salt, voices bursting out in the clamoring of those oppressed from and by salt.

Gathering research materials was a familiar chore: putting together

a bibliography for a subject—incomplete, lest it becomes paralyzing—
is part of my profession as a scientist. I thus compiled a whole file,
composed of books, offprints from various publications, travel guides,
even commercial brochures. I consulted the Internet, especially on
technical topics such as the desalination of seawater. I enlisted foreign
colleagues and friends to send me small lists of proverbs about salt in
their parts of the world. Other correspondence with heads of saltworks and with archivists-curators provided me with very precious
information on questions of a general nature, such as the salt routes,
or of a more particular sort, such as on Bayonne ham, for example.
Consulting the departmental archives of Aveyron supplied me with
precise details about the salt tax and its excesses. In short, I found
myself facing an impressive resource of research materials. How to
navigate this multidimensional space, how to organize this plainly rich
but also terribly disparate and multifaceted set of information? Since
whim was my sole guide—my bibliography was partial, I repeat, and
so as not to be influenced I carefully avoided reading works, such as
those by Philippe Meyer, Gilbert Dunoyer de Segonzac, and Robert
Multhauf, with a scope or aim comparable to mine—I imposed on
myself a rule requiring alternation. I would insert in predetermined
positions sections from three different categories of texts: prose
poems, commentaries on proverbs from various cultures, and general
reflections. I believe in literature under constraints—whether it be
Shakespeare's sonnets, the translation of Petrarch's poems (a practice I
have maintained since adolescence for my own private enjoyment, as
an enthusiast rather than a specialist of Italian), or certain texts by
George Perec or Raymond Queneau that I admire—I find it a royal
road to the appropriation of the language. These chosen constraints
gave the book its rhythm and dictated what the table of contents
would be.

I should add one more criterion that governed my efforts: respect,
threefold respect, in fact. Respect for my readers required a type of
writing suitable to a work of popularization; I had to remain simple
to spare nonspecialists a tiresome erudition. On the other hand,
respect owed to specialists on the matter, as to any reader wishing to
deepen knowledge of a particular point, led me very early on to prepare a set of notes (placed at the end of the book) that document my
sources and occasionally supply more specialized commentary. My
third deferential bow, stemming from a very sincere, deep, and long-

standing sense of respect, salutes intellectual history: I am sufficiently convinced of the worth of every bit of information, of every sign, however minute it might be, not to risk establishing or replicating hierarchies of value and value judgments. Therefore I had to welcome with equal warmth scientific publications treating of the Zeeman effect (which is the splitting of spectrum lines under the influence of a magnetic field), and of the repair of city walls in fourteenth-century Avignon, and of, further, the domination of commerce along Europe's Atlantic coast by Dutch shipping from the sixteenth century to the close of the eighteenth century. All this material formed part of the subject at hand; none of it could be left aside.

The book happened to set at that point. I use this term in the sense of the setting of a building material, a plaster or a cement that hardens, or even polymerization of monomers, observable with spectacular changes in fluidity and rheology. Crystallization offers another image, another analogy for a book's emergence into existence, as insidious as it is unexpected (though hoped for). Allow me to try to describe the happy event of the setting of a book with yet another image, even more incongruous, beginning from its opposite: nonsetting produces a result that resembles the inability of casein to form a network when rennet is added to milk, even when the milk has been boiled; one is presented not with the coveted cheese but with a rather yucky, lumpy suspension, good only to be discarded.

Authors are vain. They imagine themselves the orchestrators of a successful setting, though they are merely witnesses to the culmination—to *a* culmination, at least—of a self-structuring, one not devoid of mystery but one that I am convinced can be studied fruitfully. If an author does not give in to vanity, he or she benefits from a double awareness: an awareness of the book's strong points and an awareness of the need to work further to solidify them and to bring them to the light of clear evidence. Strong points are those aspects that appear during and as a result of the preparatory work and of which the author was unaware beforehand. The two strong points of this book are, on the one hand, the dual revelations of the callousness—to say nothing of the needless cruelty—of political power in every epoch, in our own as much as others, and, on the other hand, the extraordinarily rapid pace of collective forgetting. In fact, these two conclusions should not have surprised me, since the altogether innocent question I asked myself before undertaking work on the book was:

but how in the world did salt become such a banal product when for such a long time, or so frequently, it was a rare and costly good, "white gold," as it happened to be called?

Is it possible to draw a parallel between the art of the writer and the work of the scientist? My answer starts with the overarching concept, to be addressed in the book, of the *saugrenu*. [An adjective meaning "unexpected, bizarre, and somewhat ridiculous."—Trans.] As I shall show, this notion provides the link between the common understanding of salt and the epistemology of scientific discovery. Let us simply recall, at this point, that *saugrenu* has its etymology in the grain of salt. [*Saugrenu* is composed of *sau*, a form of *sel*, the French word for salt, and *grenu*, derived from the French term for grain, *grain*.—Trans.] To "mettre son grain de sel" is to show originality, at the least, and a genuine creativity, at the most. [The literal sense of "mettre son grain de sel" is "to add one's grain of salt."—Trans.] Indeed, scientific discovery is almost always odd and has an unseemly and inadmissible aspect when it first rears its head. But does discovery in science differ in nature from the right word arriving, mysteriously, under the pen? The right word is unexpected; it subverts and rends dominant schemas. If stereotype is the rut of thought, then the right word is a prize the imagination grants to all who adventure in writing. In the art of writing, finding a clever phrase, relaxing a sentence with a dash of humor, and making a book set as described are also kinds of miniature discoveries—*saugrenu*, at first glance—that in the best of cases may become classics thereafter.

If the creativity of the scientist making a discovery is not to be distinguished from artistic creativity, the two creators also match in a more restricted sense in terms of their routines. Writers are required to communicate. Before finding a style, a writer must write and speak everyone's language. All writing obeys norms; in order to be acceptable, a writer must remain—in a general way—syntactically and lexically correct. Likewise, scientists are never alone in their attempts to shed light on nature. They are constantly engaged in a dialogue, in a lively discussion with their colleagues, a dialogue and discussion that are pursued even in apparent solitude; researchers necessarily internalize the methodological requirements of their craft and of the small community they belong to. They spend time on various checks and controls, they carefully test hypotheses other than those they favor, and they take care to make use of adequate instruments. Likewise, they painstakingly ensure the reproducibility of their observations and measurements. Researchers show in this way that they, too, are concerned to communicate in language, in the language they share with others in the same scientific group.

I see in the two undertakings—describing nature so as to understand it and composing novel sequences of words that likewise have a new meaning—similar properties of symmetry with respect to their objects, products, and creations. What fascinates me about science, about the science that has welcomed me—namely, chemistry—is everything that pertains to the properties of symmetry in objects such as molecules, most particularly when this symmetry is accompanied by a polarity we name with the term (derived from the Greek) *chirality*. This is the property shared by a good number of molecules of being able to be assigned either to a left or a right side; here, the many logical links required to move from the microscopic scale of these entities to our macroscopic world are of little concern. What likewise fascinates me about literary texts are the echo effects that create symmetries, as it were: the instances of alliteration, assonance, and rhyme, the repetition of themes, the prosody in each language that imparts its music to a literary composition, whether it be a poem or a page of prose. On reflection, I must have written this book on salt a bit like a salt crystal is constructed, alternating between positively and negatively charged atoms. Did I not lay out the rule of alternation when I earlier made mention of literature under constraints? I've often taken a course that alternates in similar fashion, advancing an assertion to qualify it right away, evoke its contrary, or follow it with a proposition that reiterates it, but in modified form. One step forward, one step back, without ever stopping: Would this be the courage to exist? Or should one simply speak of flight in the face of the complexity of the real or of the deceptions of language?

The book follows a line, going from need and scarcity to abundance and joie de vivre; it moves from oppression to sensuality. Our Western societies have used salt for the preservation of food, in particular, to preserve proteins (chapter 1). The nomads' circuits (chapter 2) are organized not only on the basis of water sources but also on sources of salt. In addition, nomads ensure the delivery of salt to settled populations. Salt is a physiological need, to the tune of several kilograms per person per year, and so harvesting salt is essential; whether from a mine or a salt marsh, it is a highly diversified, arduous, and uncertain sort of labor (chapter 3). Political powers—and this is nearly a geographical and historical constant—arrogate to themselves control of the salt supply; they thus can organize and control its scarcity and price by means of taxes (chapter 4). All organisms, from unicellulars to humans, have a need for salt, a need that cannot be satisfied to excess without harmful consequences (chapter 5). Freed from the sole preoccupation with their survival, humans seek to know; they want to understand and master the nature surrounding them, starting with this compound of such apparent simplicity that it is at the source of scientific concepts and even of scientific disciplines such as spectroscopy (chapter 6). Finally, the consolations of myth are set against the salt of life's blood, sweat, and tears. In various cultures, these consolations range from purification to festivals, both centered on salt, and the zest it offers, imparting taste not only to food but to life as a whole (chapter 7).

salt

one **salt-cured foods**

Our very first impulse is to associate salt and the taste it imparts to a dish. And yet, its presence on the table in a more or less ornate saltcellar in a convivial setting—that of friends sharing a meal and graciously offering one another the spice—is also a legacy from history, one in which the precious spice was at times scarce and expensive. And, in yet a third meaning, salt being given and received is an indicator of the rich social relations by which outsiders integrate themselves into a gathering or, at the very least, are able to strike up an acquaintance.

The chapter thus opens with a Japanese proverb, ushering us from the world of the social contract and aggressive virility to the domestic sphere, in which the cook, whether a man or a woman, prepares dishes.

But why, after all, is salt necessary in food? Because the organism has a daily need for a certain amount of salt (at higher levels, salt becomes toxic). It is even an antiseptic agent: high concentrations of salt kill bacteria. This led to the invention of salt curing, presumably as far back as prehistoric times.

But all such inventions did not occur that early: the salting of fish (cod, her-

ring) for the purpose of preservation was invented in historic times, in the four-teenth century, assuredly, for preserving herring. Caviar, one of our luxuries, depends for its preparation on the sturgeon eggs having first been salted. In the nineteenth century, the cossacks of the Don made the preparation of caviar their specialty. An incident recounted in Alexandre Dumas's Russian travel journals hints at an elegant solution to the dilemma of how to reconcile a salt tax levied by a ruler with a communal life in which salt is to some extent demonetarized and in which what enjoys currency is the sharing of salt. This is evoked in a proverb also drawn from a Slavic culture. Salt thus symbolizes social harmony.

Having thus noted the role in food preservation still played today by salt curing, I will move from the larder to the kitchen. In cold dishes, salt is an ingre-dient essential to the taste of the food it enhances. In hot dishes, its presence in cooking water helps to protect against various denaturations, whether these are an egg bursting, pasta sticking together, or vegetables transformed into a mush not only tasteless but formless as well.

Along the way, I will show how a sauce such as the garum of Roman antiquity's common people enabled them to defraud the tax system quite as readily as to impart flavor to their dishes and even to prompt other pleasures of the flesh! Like French cuisine, Italian cuisine makes great use of sauces, such as salsa verde, and some Italian proverbs have retained their lewd double meaning, with salted dishes construed as aphrodisiacs.

The salty contrasts with the sweet. The chapter ends with the social solu-tion to this antinomy during the holiday season: it is associated both with the Saint Nicholas figure of northern and eastern countries and with the tradition of giving sweets to children; their shared origin is the salt curing of the pig, sac-rificed at Saint Nicholas tide and then macerated all winter long in coarse salt in a salting tub.

But let us first consider the outcome for other foods of being macerated for a time in salt.

THE PROVERB OF SALT ON LETTUCE.   The Japanese saying "aona ni shio," whose literal sense is "to salt the greens," means the deflating of a braggart. The word *aona* is a generic term for that which is green, more particularly, for lettuces and vegetables. *Shio* means salt. Salted lettuce wilts. It tends to shed water through osmosis in an effort to equalize salt concentrations inside and outside the plant cells (a point to which I shall return). The process is unavoidable: As soon as a droplet of water from the lettuce leaf dilutes the added salt, a brine appears. This saline solution is much more highly concentrated than that in the cells of the lettuce leaves. The two saline solutions come into contact on both sides of the cellular membranes that serve as their interfaces. Since these cellular membranes are permeable to water, the internal and external concentrations equalize, as with connected vessels where liquid levels rather quickly become the same.

But the relevant effect is the resulting decayed aspect of the lettuce leaf or, for that matter, any part of a plant. Adding salt ruins freshness, tarnishes, and makes a food less appetizing. Crisp and appealing as it was, it has turned limp and old. The Japanese proverb raises a banal, everyday observation to the symbolic level through a transfer to the moral sphere. The supposedly brave person is so only in appearance; in fact, he or she is a big coward.

Arms and the Man, an 1894 play by George Bernard Shaw, deconstructs heroism and works out in an amusing way this paradox of the warrior, who outwardly appears aggressive, martial, and bellicose yet is in fact timorous and fearful.

OSMOSIS AND SALT CURING.   Lettuce is not the only organism that wilts in the presence of added salt. To get rid of slugs, one can sprinkle them with salt. They shed their water and die. This results, for the organism as a whole, from osmosis: when water goes back and forth through cell membranes, a dilute solution will mix with a more concentrated solution on both sides of the membrane. After some time, a state of equilibrium prevails in which the concentrations have become equalized on both sides of the membrane. Recall that the concentration is the amount of substance (here, the amount of dissolved salt) per unit volume (see fig. 1).

Osmosis explains quite a few culinary practices. It is one of the reasons for salting the water used to cook an egg: if this salting is not done, the water in the pot can migrate through the porous shell into the egg's interior and dilute its content, which is richer in salt than is the cooking water. The egg would then swell up, causing its shell to burst. Another example, describing a method for treating a vegetable that closely resembles the one used to destroy slugs, is that of salting cucumbers in order to make them sweat out their water.

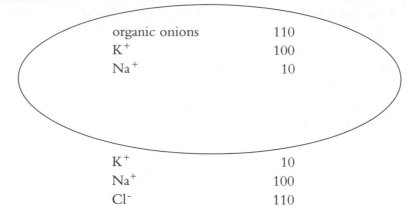

organic onions   110
K^+ 100
Na^+ 10

K^+ 10
Na^+ 100
Cl^- 110

FIGURE I. A comparison of the concentrations (in millimolar units) of the main anions and cations inside and outside a typical cell. One notes that electroneutrality obtains in each of the compartments, since the total numbers of positive and negative ions are equal. One also notices the entry of potassium into the cell and the exit of sodium. From Robert J. P. Williams and J. J. R. Frausto da Silva, *The Natural Selection of the Chemical Elements* (Oxford: Oxford University Press, Clarendon, 1996), 207.

Since organisms of all sorts, vegetable and mineral, have an aqueous interior and cells with water-permeable membranes, salt can become toxic once its concentration outside a cell exceeds that in the internal aqueous cellular environment.[1] Thus salt provides humankind with a simple technique for asepsis, as a protection against pathogenic bacteria, a technique that has been used from time immemorial, along with alcohol. On ships, sailors used it, in a painful but effective manner, to disinfect wounds.[2]

Whence, it would appear, the invention of salt curing, from the beginnings of agriculture, impelled by the need to protect various crops from spoiling, that is, from destruction by microbiological infection. As a general rule, salt-curing methods go hand in hand with partial drying techniques, aimed at preserving protein in a more lasting way:[3] milk transformed into cheese, salted fish (most particularly, herring and cod),[4] salted meats (dried meat from Grisons or Italian *bresaola*, ham, dry sausage, various *charcuteries* [cooked pork meats]).

The making of ham is an example.[5] It is done in December and January because the hams must be prepared in cold weather (at a temperature below 39.2°F or 4°C) that will last long enough (thirty to forty days, or, to be accurate, two and a half days per half-kilogram) so that they won't spoil before the salt curing can protect them. This period of time explains the connection in Western nations between the Saint Nicholas festival and the period of the year for slaughtering pigs to prepare salted meats, which serve as a repository for protein. I will return to this.

Typically, the processing is done with a mixture of salt, sugar, and saltpeter. Sugar serves to counteract the salt taste and supply energy to the bacteria that transform nitrates into nitrites. Among other functions, saltpeter acts to redden the meat, which would otherwise be a not very appetizing gray color, retard rancidity, and prevent botulinic toxins from developing.

But though the salt curing of a ham occurs because salt protects it from external bacteria, the process also makes use of internal microorganisms: *Micrococcus auriantiacus* transforms sodium nitrate into sodium nitrite, gluconodeltalactone converts sodium nitrite into nitrous acid, and the ascorbates then free nitric oxide, NO. This reacts with the myoglobin in the meat to produce, in an irreversible process, a compound called nitrosyl hemochrome. If necessary, the salt-curing process is then followed by others, such as those used to make smoked ham.

SALTING HERRING. Salt curing ensured the preservation of herring. To do it, one slits the fish with a special knife and removes the gills and the branchia, the heart, and part of the viscera.[6] The blood empties from this wound. Pancreatic enzymes, remaining active in the body of the fish, partly digest its flesh and make it tender. The herrings are then packed together with salt in a barrel (a *caque*, "cask"—hence the name of the process. [In French, the salt curing of herring is named, following the original Dutch term, "le caquage des harengs," the title for this section of the chapter in the French original.—Trans.]

The procedure appeared early in the fourteenth century (in Flanders circa 1315–1330).[7] A semilegendary character, Willem Beukelsz (or Beukels, Beukelszoon, or William Benkelsoor), is generally credited with its invention.[8] A fisherman or steersman from Biervliet in Zeeland, he died in 1397, if we are to believe a stained-glass window in the church in the town of his birth, which shows him in the process of salting herring. The emperor Charles V visited Biervliet on August

30, 1586, and honored his memory on that occasion. But the very dates of Beukelsz's life are disputed. He allegedly was the deputy magistrate of his village in 1312, but some historians date his discovery to 1384; according to others, who date his death to 1397, the discovery was made in 1349; and according to still others the discovery dates from 1375.[9]

Whatever the historical truth, this astute way of combining salt preservation with the beginnings of a digestion process using proteolytic enzymes was a revolutionary technique.[10] It would be responsible for the prosperity of fishing centers such as Aberdeen;[11] the Shetland Islands, to the north of Scotland;[12] the Hanseatic cities specializing in the twin trades of salt and herring; and the Netherlands especially: according to a proverb, Amsterdam was built on herring casks.

THE COSSACKS OF THE DON.   Caviar is another salt-cured food. It consists of sturgeon eggs preserved with salt. In the nineteenth century, the cossacks of the Don had a near monopoly on its production, which they used as leverage to exact from the czar noninterference in their traditional way of life. An observation from one of the greatest French writers intrigued me and led me to discover this cunning tactic used by the Cossacks.

Toward the end of his journey to Russia in 1858, Alexandre Dumas again met with those cossacks of the Don who, as he wrote, "gave us such a great fright in our youth" [presumably at the time of the invasion of Paris, after Napoleon's final defeat]. He wonders about their resources:

> They pay the costs of their upkeep themselves [...] common soldiers receive [...] only thirteen rubles a month. With these thirteen rubles, they must clothe themselves and supply their horse and weapons. [...] They make do as best they can. It is up to them to get through hard times *without sin*. Russia is indeed the land of impossible arithmetic problems.[13]

Instead of yielding to this absurdity, Dumas would have done better to suspect that he lacked a piece of information. Puzzled by his remark, I was able to find the answer to it in the travel account of the Westphalian baron August von Haxthausen, published in 1847.[14]

The cossacks of the Don lived in a community governed by strict rules, which actually brings to mind the kolkhozes of the post-1917 Communist regime. The cossacks owed military service to the czar, but the wealthiest of them could buy themselves a replacement, which led to some income redistribution among families. However, the central unit of cossack economic life was not the family but the entire community, what one might term the cossack nation. Two chief sources of income prevailed in its budget: the sale of sturgeon-fishing permits and a salt tax (salt was indispensable for curing of the fish and of the fish roe, or caviar). The sale price of a single sturgeon could reach as much as four hundred rubles, the yearly income of the average cossack.

Alexandre Dumas committed two errors, then, errors of ignorance or oversight: the thirteen rubles from the government were in fact just pocket money for a cossack, who lived on the community as a whole rather than at his own family's expense. Moreover, the cossack community—one is tempted to write "the cossack commune"—had plainly negotiated a treaty with the czar to supply him with elite soldiers in exchange for the right to administer salt taxes. The cossacks occupied an area of southern Russia, between the Don and the Volga rivers, a region that, though far from the Caspian sea ("three or four hundred versts away," according to Dumas, or about four hundred kilometers [A verst is a Russian measure of length, 1067 meters or two-thirds of a mile.—Trans.]), nevertheless abounds in salt lakes that, to use Dumas's description, "yield . . . fourteen to fifteen million kilograms of salt annually."

Thus, in sharing revenues from such saltworks with the government, the cossacks freed themselves from dependence on the merchants who bought their fish and otherwise might have supplied them with salt at exploitative prices. Moreover, the communal structure of their economic life, strictly egalitarian in a great many respects and ordered by stringent rules, prevented any given family from acquiring a monopoly on the provision of salt.

One can contrast this good fortune of the cossacks of the Don, which they owed to their favorable geographic location next to their own salt supply, with the less bountiful fate of numerous fishing peoples. The cod-fishing people of the Shetland Isles in the north of Scotland were supplied with salt, alcohol, and tobacco by none other than the buyers of their fish, who arrived from the

Hanseatic cities of Hamburg, Bremen, and Lübeck. The ensuing near-colonial domination of these buyers would last from the fourteenth century until the beginning of the eighteenth century, well after the decline of the Hanseatic League in the sixteenth century. Finally, the union of England and Scotland in 1707 led to the levying of taxes to dissuade foreigners from trading on British soil. Essentials: scarcity of one commodity (salt) can lead to abundance of another (friendship).

THE PROVERB OF FRIENDSHIP OVER SALT.   The social importance of the meal, the conviviality that accompanies it, in a warm and friendly atmosphere, have meaning only in an economic context of scarcity, where abundance is something of an exception, where being able to eat one's fill is cause for celebration, where, even if harvests have been good, shortages and famine remain genuine threats. Over the course of centuries, most of humankind has been steeped in such stark tradition.

A Polish proverb says "zjesc z kims beczke soli." Its literal translation is "to have eaten a cask of salt with someone." Thus it carries the meaning of a deep and enduring friendship, one nourished on the relations enjoyed by longtime table companions.

To share bread and salt was and remains the symbolic gesture of hospitality, of the welcome offered a stranger. Familiarity implies the regular repetition of this gesture. Since salt, this costly, indispensable commodity, is nonetheless in the end consumed in substantial amounts—whether a *minot*, a bushel or, as in this case, a cask or barrel—the fact that one would have shared a great quantity of it comes to signify a lasting friendship.

The fact that volume units rather than weight units are used to measure amounts of salt bought or sold most probably relates to its means of production in the saltworks: salt is gathered, like a grain; it thus admits of the same units of measure, that is, the volume measure of dry granular, or powdery foodstuffs such as wheat, oats, barley, or even lentils. Salt will thus be measured by the *muid* (cask), *feuillette* (half-cask), *quartaut* (quarter-cask), *velte, pot, pinte, sétier, demi-sétier, posson*, and *roquille* (one *muid* is equivalent to 288 pints, or 1,152 *sétiers*; the *quartaut* is one-quarter of a *muid*; the *velte* is 16 *sétiers*, and the *sétier* is 4 *possons* and 16 *roquilles*).[15]

FOOD PRESERVATION.   These ancien régime units of measure have become obsolete. But salt-cured foods are still around. When shopping at the supermarket, we are unlikely to register that we are visiting a museum of technology. But, unlike other sectors of the economy—in sound reproduction, compact discs and DVDs have hardly left any room for LPs and 78s—the food trade is conservative!

To be found side by side on your grocer's shelves are dry sausages and other salt-cured foods whose origin goes back at least to the Middle Ages, canned goods (invented by Nicholas Appert between 1795 and 1810), refrigerated packaged goods (though Romans already used refrigeration, industrial refrigeration was developed during the course of the nineteenth century), pasteurized foods, such as milk, beer, and numerous cheeses (the technique invented by Pasteur dates back to the 1880s), extracts of meat broth (another chemist, Justus von Liebig, reinvented Lavoisier's method of dehydration in the first half of the nineteenth century), dried fruits (dating back to antiquity), frozen or freeze-dried foods (both techniques date from the second half of the twentieth century), condensed milk (invented by Gail Borden in 1856), and more.

All these methods have in common the sterilization of foods (pasteurization) or at the very least the slowing of bacterial proliferation (frozen foods). It is also often important to deactivate certain enzymes, those present in meat, for example. In fact, proteins are the ingredients made inactive by subjecting food to a high or low temperature, by changing the acidity of its surrounding environment (pickles or onions in vinegar), or by salting it. High concentrations of salt (cod or ham) or of sugar (jams) are toxic for many microorganisms.

Spoiled foods can be dangerous because of the multiplication of infectious microorganisms such as *E. coli* and *Clostridium botulinum*, which produce highly poisonous toxins. Keeping the temperature low enough (below -25°C) prevents this last bacillus from multiplying and secreting botulin, one of the most toxic substances known.

We owe a certain number of food preservation techniques to the military. In 1795 France, the Directoire—the government at the time—offered a prize to the inventor of an effective method of food preservation. After a good many attempts, Appert recognized that heating must be combined with excluding air from a hermetically sealed container. Napoleon awarded him the prize in 1810 after the

French Navy confirmed that Appert's rations survived 130 days at sea without spoiling.

Across the Channel, from 1814 on, the British Army and Navy also supplied their men based overseas with canned goods: in 1810 Peter Durand received a patent from George III for his canning technique, which was developed and marketed by Bryan Donkin and John Hall.

Canned food, which played a large role during the California Gold Rush of 1849, was essential to the Civil War. In the twentieth century, during the First World War, the daily rations of soldiers at the front consisted of the following: for the Germans, either 375 grams of fresh meat or 200 grams of canned meat (and 25 grams of salt); and for the English Tommies, 560 grams of fresh meat or 460 grams of canned meat (or salt meat), with 14 grams of salt. The can of "bully beef," or corned beef, as we know it, was invented in Chicago in 1875 by Arthur A. Libby and William J. Wilson.

In 1943 the American army perfected a technique of food sterilization using irradiation by a radioactive source. Despite its development and marketing in forty or so countries, this process runs into the phobia aroused in public opinion by all things nuclear.

Will other technologies see the light of day? Without a doubt. Some of them are intellectually attractive, such as pasteurization by means of pulsed electric fields.[16] This method is somewhat akin to the ultra-high-temperature sterilization (UHT) of milk, in which the foodstuff is heated for several seconds to about 120°C. In this case, pulses from an intense electric field over several microseconds annihilate the bacteria, for the microorganisms are destroyed after their protective membranes are perforated.

One could also resort to bacteriological warfare, but would it be any better received than irradiation has been? The admittedly handsome idea is to destroy harmful germs with other microorganisms that would become their predators while themselves remaining benign.[17]

Perhaps the greatest promise for research in food preservation is in some of its fringe benefits: the use of freeze-drying techniques in the conservation of books and old documents or their application by archaeologists to wooden ships raised after long periods underwater. Biochemistry is quite naturally a field ripe for the application of all these techniques. Our modern larders thus form a sort of exhibition gallery for the technologies of every age, radiating out to other realms.

FLAVOR CONCENTRATES.   As mentioned, food refrigeration is quite ancient. Yet we continue to feel partial to foods preserved by drying and salting. They range from dry cheeses to dried meats, from country ham to dry sausages, and likewise include dehydrated broths such as beef bouillon cubes and fish such as salt cod. If one adds to this list corned beef,[18] caviar, rollmops, smoked herring, and other sausages and *andouillettes* [An *andouillette* is a small pork tripe sausage.—Trans.], that is, salted but only partly dehydrated foods, one finds the category of salty protein to be large and diverse.

Why have so many such foods persisted into the age of refrigeration? The obvious answer, given the conservatism inherent in every culture and given also the psychological association between such lightweight nourishments and outdoor adventure, is indeed their saltiness. Combined with dryness, it can make quite a bouquet of flavors blossom.

On the one hand, sodium cations (atoms positively charged through loss of an electron) bring to the tongue's taste receptors not only chloride anions (atoms negatively charged through gain of an electron) but also other sapid anions. On the other hand—this is just a conjecture of mine—proteins impregnated with sodium from salting, because they are made of twenty different types of amino acids, contain here and there glutamic acid (one of these amino acids). Thus they liberate— through hydrolysis during digestion—sodium glutamate, either directly or as peptides (linking a small number of amino acids) locally containing a glutamate unit. Many of us are familiar with flavors enhanced by monosodium glutamate, a spice in Chinese restaurants.[19]

Well before Marco Polo, when the West began to experience Chinese cuisine, the enhancing of flavors with salt was a given in Mediterranean cuisine. Sauces partake of the same concern to make dining even more enjoyable.

SAUCING.   The Romans seasoned their dishes with a sauce they called *garum*.[20] Pliny mentions it, as Apicius does, in his cookbook.[21] *Garum*, to give it a first rough description, consisted of spoiled fish mixed into a brine.

Let me further refine the picture: *garum* was prepared by crushing together, with a small amount of salt water and/or wine or olive oil, the flesh of various fish, to which one also added their entrails; the digestive enzymes in the fish caused a partial digestion, and the chief biological role of the salt was to make the mixture aseptic and thus to prevent external bacterial germs from proliferating.[22] In this way, one concocted a very salty condiment with a powerful, practically revolting smell and a taste that must have somewhat resembled our anchovy pastes.[23] At meals, the Romans could not do without it. It is likely, too, that the common people of Rome used *garum* to evade the salt tax.[24]

A recently re-created menu for a Roman dinner opened with, as an hors-d'oeuvre (*gustatio*), bean salad (made with black-eyed peas) with coriander, raw leeks, cumin, and *garum*; salad of pork and veal trotters and snouts seasoned with pepper, celery seed, and *garum*. There followed, as an entrée, or *prima mensa*, roast pork with pepper, honey, vinegar, and *garum* and smoked pork and beef sausages with pepper, cumin, savory, rue, parsley, bay leaves, pine nuts, and *garum*.[25] In various countries, such as Turkey, versions of *garum* survive to this day.[26]

One can define *garum* as a sauce, and indeed the word *sauce* derives from the Latin adjective *salsus* (salty). A great number of terms in the various Romance languages also have *sal* (salt) as a root.

Such lexical richness is mirrored in the great variety of sauces that the ingenuity of male and female cooks has devised over the course of centuries. If *garum* (the *garon* of the Greeks) is one of the recipes bequeathed to us by the classical age, the fact that it is a salted sauce, often intended for the poor, has been preserved in cooking terminology and in cookbooks through all the "poor man's sauces" found there. One of these is the "sálso del paure houóme" from Aveyron,[27] made most often of wine and bread, for which *Cuisinier Royal et Bourgeois* gave this recipe in 1721: "Take some chives, skin them well, and chop them up nicely; once chopped, put them in a sauceboat with some pepper, salt, and water and serve cold."[28]

That bread crumbs would serve as a binder for the wine, salted to make a sauce out of it, has left a trace in the vocabulary: in the Occitan language spoken in southern France, bread soaked in wine is called *saussole, saoussole*, or *chaouchole* ("*saucette*" [little sauce], in a way),[29] with the variable spelling reflecting differing pronunciations.

On reflection, lexical richness and the abundance of sauces have very similar origins. A word is composed of a root and a suffix (to take the above case, *saussoira* is composed of the Latin *salsus* plus *oria*), while a sauce is composed of a flavoring and a thickening agent.[30]

Which brings up spices. Rather, allow me next to spice up the narrative with a personal confession. It doesn't concern an actual misdeed, only a strong temptation.

SAUMANDISES.   Throughout the writing of this book, I had thought of entitling it *A Short Treatise on Salt*. It thus presented itself to me, in imagination, as a work of nonfiction. Despite this, I have had periodic wishes to slip into fiction. Since a treatise aims to be, and is called, scholarly, supporting each one of its claims and each fact cited with a real or assumed bibliographic reference, I fantasized about shifting this framework toward the imaginary side of things.

Let me sketch out a typology of *saumandises* under three categories. A *saumandise* could be a precise fact endowed with a sham reference and would thus remain unlocatable: that's the first kind.

Another way of pulling back from the overseriousness of a wholly and thoroughly documented handbook consists, conversely, in putting forward an imagined fact but accompanying it with a scholarly foundation, with that library humus one calls a bibliography, an authentic but meandering one in this case.

A third possibility for trickery in nonfiction writing presents a wholly fictional fact rooted in referential compost that is likewise a product of sheer imagination.

I would call these various attempts at miswriting—where miswriting is to writing as the historical novel is to history—my *saumandises*. Here is what one could read in a dictionary under the entry for *saumandise*: feminine noun—(Medieval Latin: *salmans, salmandis*) 1. on the face, trace of the path of tears. 2. Encycl. viscous trace left on the ground or on a plant by a slug or snail. 3. Bot. salt-rich sap of marine plants such as saltwort. 4. Lit. akin to the aphorism, this figure of style (or trope) expresses sober truths through humor. The novelist Alexandre Viallatte was expert at writing *saumandises*.

It remains for me to offer up three *saumandises* to illustrate their various garb. The third kind of *saumandise*, a tall tale to be precise, would be the following: Jacques Loeb (1859–1924), while investigating gelatin at the very end of his career, discovered that the amount dissolved in a given amount of water increased in proportion to the saltiness of the liquid.

A *saumandise* of the first kind would be, for example: we are indebted to the Greek geographer Strabon, who lived in the first century before Christ, for the first description of the saltworks at Guérande.

Finally, a *saumandise* of the second kind would be: In the 1880s, Swedish pharmacies sold hemoglobin tablets, reputed as revitalizing. The active ingredient was hemoglobin crystallized by salting of the red corpuscles, according to the method perfected by Félix Hoppe-Seyler (1825–1895).

The above *saumandises* are the only ones, to my knowledge, that appear in this book. Still, I regret my timidity for not having hidden others, here and there, the way one places little presents in the yard for the children on Easter Sunday, a treasure hunt allegedly initiated by rabbits.

The name *saumandise* brings that tradition to mind. *Saumandises* have the merit of imparting more flavor to a text. They are the conversational equivalent of the joke, the verbal exaggeration, the Anglo-American (or southern French) tall tale.

These pinches of salt, some claim, would prevent the ink from drying completely. Their parodic nature compels authors to reflect on the aims and reach of their texts, since it is so clear that readers hear what they will, sometimes hearing in them the echo of truth (or of a truth) even more clearly than when that truth was expressed initially, devoid of all fancy.

In the end, shouldn't all writing practice self-mockery and retain a distance from its obvious assertions, a distance established through humor and marked out with the salt of jesting and tall tales, as in conversation, at the dinner table or in the kitchen?

COOKERY.   Appetizers and snacks can serve as cultural indicators. Raw foods spiced with just a sprinkling of salt prefigure cookery. Undoubtedly, this is why we begin (or precede) our meals this way.[31] To put a bit of salt on a tomato, radish, pepper, stalk of celery, or slice of fennel or carrot, besides combining two flavors, carries with it a strong symbolic meaning: before ingesting this food, we present it to the gods, in a distant remembrance of ritual sacrifices. In other words, we transfer food from nature, where we culled it, to culture. Eating raw foods with a few grains of salt is an allegory for the cooking of food, of which it also forms the very beginning.

All natural foodstuffs undergo a double denaturation, usually though prior domestication of the vegetable or animal species (with the three great exceptions being the booty acquired from hunting, fishing, and foraging) followed by cooking. In this way, we promote a natural product to culture; we culturize it, to use a neologism.

This culturalization obeys strict rules, but this is not the place to set them out. First, one must note that taste requires educating. In my French culture, most very young children do not spontaneously like olives or oysters; we gradually raise them from a sweet diet toward a gastronomy of savory foods. Further, cultures differ from each other in their tastes in food: the French have little fondness for the Scotch *haggis* or the Japanese *natto*; Americans refuse our *andouillette*, calf sweetbreads, and tripe, and wild mushrooms terrify them; the inhabitants of Central Europe are often hesitant about seafood, and so forth. These aversions bear on the two major qualities of a dish—once in the mouth—its consistency and its taste.

Having recalled this truism, the cultural embeddedness of all cuisine, let me venture this statement: every culturalization of a food occurs through violence against nature. The subjective aspect here is just as important as its objective reason, namely, the food's detoxification, the annihilation of any potentially toxic microorganisms, which are rendered harmless through elevated temperatures (boiling, cooking for often long periods of time), or by means of a highly acid or base medium, or through the use of salt.

Culturalization is often a cold sort of violence, preceding the hot violence of cooking: squid pounded so as to tenderize them, pasta passed through rollers, ingredients pulverized in a mortar, eggs or cream beaten or whipped (the verbs are significant). Moreover, the very verb *cuisiner* (to cook) also carries the secondary sense, applied to a person, of subjecting him to interrogation, with the connotation of torture intended to make him confess his crime. [The figurative sense of *cuisiner* is close to the figurative sense of the English "to grill," meaning "to subject to severe questioning."—Trans.]

One prepares something for eating, and the verb *faire* (to make) has here the strong, objective sense of triggering a physicochemical process that is often highly complex, though sometimes susceptible to rational understanding, as well as the subjective meaning of a voluntary act and a procedure one follows with deliberate intent. The syntax of recipes is revealing in this regard: *faire* revenir, *faire* bouillir, *faire* dégorger, *faire* mariner, *faire* tremper. [These are the equivalents of "to brown or fry," "to bring to the boil," "to leave to sweat," "to let marinate," "to soak or dip." French and English syntax differs in these sorts of expressions. The literal translations of the French expressions would be: "to *make* brown," "to *make* boil," "to *make* sweat," "to *make* marinate," "to *make* soak." The French expressions, with their evocation of a subject as a causal agent, demonstrate the author's point concerning the subjective aspect of the making that works to culturize foods, a point not discernible in their standard, idiomatic English equivalents.—Trans.]

For cooking is a serious business, not child's play. The cookery recipe is a set of directives that call for precise, meticulously timed actions and movements. The production of a dish is ritualized in this way, as is its collective consumption as a meal. Adding salt when producing or consuming a dish—a gesture that attains the height of ordinariness in the pinch of salt added to the pot or when one reaches out to help oneself to the salt at the table—exceeds its primary function of imparting a desirable flavor to food: it is also a sign of the enduring nature of a ritual of culturalization.

What, then, is a meal eaten in a group if not also a ritual of collective and deferred consumption? One recalls the Luis Buñuel film *The Phantom of Liberty*, with its provocative scene that inverts norms and taboos: there feeding oneself was depicted as an intimate and solitary act, whereas defecating was done in a group. The example illustrates well the social importance of meals. Moreover, because it permits food preservation, salt in itself typifies a meal that is inherently a deferred consumption of food, since it is part of the stored harvests and reserve stocks of a human group.

Adding salt, contrary to our naïve idea of it, affects not only a food's taste but also its texture. It is easy to convince oneself of this if one forgets to salt the cooking water when preparing pasta or if one prepares crêpes Suzette with a sweetened batter that has not also been salted.

Cooking obeys the epistemological model of empirical discovery and falls within the province of combinatorial chemistry. The number of possible combinations is very high. A chef trained in a catering school followed by apprenticeship can devise billions of different dishes. Any one of us in fact has a repertory—a potential one, of course—of free-form dishes (as opposed to the forms imposed by standard recipes) on the order of one million. How do I come up with such a figure? By multiplying the number of different meats available (twenty or so),[32] by the number of fruit or vegetable accompaniments (about twenty),[33] by the number of sauces they can be served with (about ten),[34] by the number of flavorings one can impart with spices (also about ten)[35] and/or herbs (about half a dozen),[36] and, finally, by the number of cooking methods (five distinct types).[37] This produces a total of 1,200,000 different dishes—not all delicious, of course.

But let's give a concrete example: (pork-duck-pollack) multiplied by (oranges-turnips-sorrel). Nine meat-vegetable combinations are identified in this way. Although I have never yet tasted fish with oranges or duck with sorrel, these combinations seem interesting and worth trying. But inventing a new recipe does not quite match the elation of falling in love, does it?

THE PROVERB ON SUCCESS IN LOVE.   The Abruzzi proverb "l'amore né vvó le bbellézze; l'appetite né vvó la salze" literally means "love does not want beauty; the appetite does not want sauce." My first interpretation is that this should be understood to mean: "as one is not sated by a sauce alone, requiring something more substantial, so love is not fulfilled by the beauty of the beloved." Here is another: "as a sauce enriches a dish, so in love beauty is nothing but a possible garnish; it comes as something extra." Thus the proverb functions as an aphorism with a moral content.

The parallel it establishes between love relations and cookery is not unusual. Gastronomic metaphors are common in matters of love. Among them we find "being drunk with kisses" and the "consuming passion."

This proverb, as is common in this poetic form, imposes a strict order on its phonetic field. The symmetry between its two parts is clear. Both halves open with an assonance (l'a) and close on a rhyme (ze). The conciseness of this proverb is its strength. French requires that it be spelled out in greater detail than is necessary in the Abruzzi dialect. Translating, here, is a tricky matter.

"Love can do without beauty, the appetite without sauce" is a viable rewriting. Every version of the form "As the appetite is not satisfied with a sauce, so beauty is not enough for love" distorts the structure of the original. To translate while seeking to preserve both the assonance of the clauses and the rhyme seems impossible here. "Fine will be your fare if your lass is not fair" seems an acceptable variant.

But let's go from the sauce back to cookery, to examine the relation of raw, salted food—and salting—to cooked food. This relation is established by means of a utensil more from the larder or the cellar than the kitchen, namely, the salting tub.

THE SALTING TUB.   In a sense, the salting tub—a stoneware vessel as ample as the belly of a pregnant woman, able to hold as much as a whole ham or a haunch of wild boar in brine, while remaining portable enough to be lifted, carried up from the cellar, set on a shelf, cleaned, and restocked with a new batch of provisions—whose function was to conserve and preserve its contents, was the ancestor of the refrigerator.

For millennia, potters have created forms thrown on a wheel, thus axially symmetrical, that provide the most volume while occupying the least space and so are well suited to storage; these forms hence approach the fullest form possible: the sphere. The amphora and the funerary urn are two variants of such vessels and represent material compromises between cylindrical and spherical forms.

The salting tub, whose shape has remained unchanged in the Western world for centuries on end, is another example of this, a pot form so well designed that it is pot-bellied. Pottery, a craft akin to cooperage and basket making, like them is several millennia old and resembles them also in its utility for daily life, as well as for survival.

In all seasons, whether of mist or frost, of shortage or famine, of Lent or Christmas, of youth or old age, of engagements or weddings, of baptism or burial, of sickness or epidemic, of cold or drought, of harvest or grape picking, the salting tub was the first and final resource. This modest household item, one of our forebears' larders for the preservation of proteins, was the anonymous or collective invention of a fertile, pragmatic imagination: a coffin, ample as a pregnant woman's belly, it held dead flesh, sheltering it in the service of the living. The Legend of St. Nicholas salted away and preserved in the collective unconscious the essential notion: keeping flesh fresh in a salting tub.

Do children believe in Father Christmas, the French Père Noël? The Anglo-Saxon avatar of Father Christmas is called Santa Claus, the translation of Saint Nicholas.[38] In northern and eastern France, as well as in Central Europe, Saint Nicholas is the titular saint and protector of children, the one who brings them playthings and sweets on his feast day, the sixth of December.[39]

Until recently, Saint Nicholas was accompanied in France by his opposite, Père Fouettard (literally, Father Whipper). Saint Nicholas rewarded the good children; Père Fouettard punished the bad ones, giving them the whip.

A story explains the reward linked with Saint Nicholas. In this tale, three students, having got lost, seek shelter for the night from a butcher. The butcher kills them, cuts them up into pieces, and puts them in a salting tub with some pork.[40] But Saint Nicholas manages to recover the remains of the three vanished young men and to resuscitate them. Since then, Saint Nicholas has become the patron saint first of schoolchildren and then of children in general. In order to sketch out a structural analysis of the tale, I will first draw out various connections from the study of folklore.[41] The feast of Saint Nicholas takes place at the start of the winter season, at a time of year when, in the countryside, fieldwork is prevented by bad weather. Traditionally, it was a period when young people were idle. It was a custom—one that traces back to the Roman Lupercalias—that, in a festival that was a genuine winter carnival, the young men pursued the young women; they were armed with whips whose lashes were supposed to sexually arouse their prey.[42] The onset of winter was also considered to be the period when the dead return to earth, perhaps to perform their dance of death.[43]

We come now to salt.[44] The three children rescued by Saint Nicholas were put in the salting tub, along with quarters of pork. Indeed, it was in winter when, in the countryside, the family pig was sacrificed, then salt cured from Saint Nicholas's feast day until Mardi Gras.[45] What is more, Saint Nicholas is the titular saint of Lorraine, a region known for its salt reserves.[46]

Let me try to tie the disparate elements of the myth together into a coherent structure. One obvious starting point is the good-bad dualism, represented by the pair consisting of Père Fouettard and Saint Nicholas. It prompts one to seek out other equally stark dualisms. The legend of Saint Nicholas presents us with several others: life–death, pleasant season–dreary season, animal (pig)–human (children), and salty–sweet.

In the same way that the dead return to earth, the students are resuscitated, and the carnivalesque bacchanalia to which I alluded unfurls with something of an excess of vitality in a ritual of animal fertility. The salt curing of pork is performed during the dreary season; it runs from the beginning to the end of winter, joining winter to spring. The salting tub, a vessel swollen like the belly of a pregnant woman, recalling the uterine nest, is likewise, in this legend of Saint Nicholas, the hyphen between the animal and the human. Another way of putting this is to note that, with respect to its container and duration, the salt curing of pork is a little bit like a gestation or a germination: a lethargy, a seeming death, precedes and prepares the edible meat. Finally, salt contrasts with sugar, and we know the place of sweets in the festivals of Christmas and Saint Nicholas. It is as if an extravagant consumption of sugar, a potlatch of sugar consumed in great excess, were a response to the outlay of salt: salt was a precious commodity, and the salt curing of a pig used a vast amount of it.

two **nomads**

Settled and nomadic peoples share an essential need for the nourishment salt. For nomads, carefully predicting the load of salt that will be needed is essential to survival. In the desert, they thus maintain their routes in order to guarantee their supply of salt. The camel is the pack animal that humans domesticated to carry the precious commodity, importing the beasts from Arabia.

A desert landscape presents a great profusion of colors and textures even to the unpracticed eye. It displays a studded, glinting gallery of minerals with numerous salt outcrops, most of which are the remains of salt lakes (or evaporites). A reference to them dates as far back as The Histories *of Herodotus.*

Saint-John Perse made himself the poet of the forlorn individual in such a richly bleak landscape in the midst of haughty, mineralized nature. Is this image a metaphor for modernity, as it was for Baudelaire, or is it an escape from the degrading mediocrity of society, as it was for Jünger?

But Earth is not made of desert alone. Temperate areas have their salt routes, too. Traces of such routes can be found in archives, and certain paths

still remain under other, less obvious guises. For humans and animals can-
not do without salt, as traditional sayings confirm. In livestock-rearing areas,
salt licks are provided in the pastures and during transhumance, a relic, in
our settled lands, of nomad life.

Many locales have long been devoted to salt production, as geographic
place names indicate. An entire technical terminology also provides evidence
that the major salt production sites in France—the salt marshes of the
Mediterranean and the Atlantic, as well as the saltworks of Franche-Comté
and Lorraine, which used igneous production methods (heating brines to
evaporate the water)—were kept in technical isolation from each other over
the course of their histories.

SETTLED AND NOMADIC PEOPLES. Settled and nomadic populations oppose one another in every manner. Even today, when
nomads have all but disappeared from the globe, when what remains
of them are merely tattered, shrunken populations, as with the Gyp-
sies of Western Europe or the African Tuaregs, traces of this opposition persist. If we are to believe Lévi-Strauss, one's personal preference to wash in running water, under a tap, or in standing water,
from a filled basin, attests to an inherited memory of this age-old
cultural dualism.

The book focuses on salt, a chemical that to us has become very
mundane. Settled populations have always needed a supply of salt. In
fact, this is why archaeologists locate the first human settlements along
sea shores, where extraction of the sea's salt can be done by simple evaporation in the sun. The later inland migrations of these first settlers were
made possible only by the specialized labor of coastal dwellers who produced salt and then sent it along trails to those living inland.

As for nomads, they must accurately estimate how much salt to carry
with them on their travels; they must draw up maps of their routes,
indicating not only water sources but also the locations of salt deposits,
of outcrops on which they will be able to draw, or even of caches or
stocks already established. If they load up with too much salt, this will
weigh them down. If they don't take along enough to see them to the
next source, both humans and animals will feel the lack severely: the
nomads will risk death for both themselves and their caravans.

And we may continue to witness such behavior, even today, by
observing nomadic desert peoples.

ON CAMELBACK.   From very ancient times, pack animals have ferried salt across the Sahara desert. In the fifth century before the Common Era, Herodotus had already drawn the connection between salt resources and the human settlements of North Africa:

> Up country further to the south lies the region where wild beasts are found, and beyond that there is a great belt of sand, stretching from Thebes in Egypt to the Pillars of Heracles. Along this belt, separated from one another by about ten days' journey, are little hills formed of lumps of salt, and from the top of each gushes a spring of cold, sweet water. Men live in the neighborhood of these springs. . . . The first of them, ten days' journey from Thebes, are the Ammonians, then come the Nasamonians at Augila with its date-palms . . . the land of the Garamantes [the Sahara] . . . the Atarantes of Mount Atlas . . . and finally the Atalantes.[1]

The salt deposits Herodotus mentioned were most certainly evaporites that had resulted from gradual desiccation of a body of salt water.[2]

Sizable camel caravans cross the Sahara desert with salt cargoes even today, following a number of different itineraries.[3] These *azalaï*—to designate them by their name in Tamahaq, the Tuareg language[4]—boast hundreds if not thousands of pack animals, as documented in *The Camel*.[5] One of the chief reasons for the great size of the convoys is security against possible attacks (the same motivation that prompted the use of convoys in the North Atlantic during the Second World War). The caravans pay a right of passage (122) to the tribes whose territories they traverse, in return for protection. As a general rule, each pack animal carries four cakes of salt, or 140 kilograms (161).

The camel was introduced into Saharan Africa with the Arab expansion, in the seventh century.[6] In the previous century, Muhammad had discovered the formula for a conquering Arabism, for a nation founded on religion, in order to fight the double yoke under which Arabia then found itself: that of Byzantium and Persia. Caravans, but of horses drawing carts[7] (we know this from paintings of the Tassili of Ajjer), already crossed the desert at that time. The hardier and more profitable (118–20) camel then replaced the horse (ch. 5).

From the twelfth to around the sixteenth century, gold arrived in North Africa from the great kingdoms of Black Africa (Ghana, Mali, Songhrai) (121). Timbuktu, on the arc of the Niger River, was the great trade center of Black Africa. In the other direction, from north to south, the main trade routes went from the south of Morocco to the Niger, either by way of Mauritania or following the valley of the Saoura wadi, and from Libya to Chad by way of the Fezzan (122).

The traffic in salt, still going on today, is the sole remaining trace of the teeming caravans that in centuries past also conveyed gold and slaves between Libya and Niger. Salt—as essential to the life of humans as it is to their pastoral way of life—is worth gold in regions that lack it. In fact, in the twelfth century salt from Sidjilmassa in southern Morocco was traded in Ghana for its weight in gold. In times nearer our own, one slave was traded for a block of salt the size of his or her foot (122).

Typically, these caravans pursue their course during the six winter months, and the camels are rested during the rest of the year. One circuit, going from the rock salt deposits of Mali's Taoudeni to Timbuktu, 700 kilometers to the south in that nation, takes three weeks at best (161). Most authors on the subject mention a caravan that got lost on the return trip from Timbuktu to Taoudeni in 1805: two thousand people and eighteen hundred camels died of thirst (122). Bilma, in Niger, also has outcrops of unearthed salt. It is the departure point for caravans going to Agadez (162), also in Niger, though more to the south: for five solid days this route crosses a zone of utter desert, so the camels must carry their own feed.

Even today salt transport by camel is less costly than transport by truck. Bilma produces four thousand tons of salt per year; twenty-eight thousand camels are needed to move it. Each animal is estimated to be good for four years of effective use,[8] so the camel fleet has to be replenished at a rate of seven thousand head per year. This yields a return of two million dollars for the breeders. The political powers that be maintain both this traditional order, on which they depend, and the tribal structure of the desert nomads by outlawing transport by truck (159). Are they being smart? Indeed, salt is a cognate of wit.

MIND OF SALT.   [The connection between the senses of *sel* (salt) and *esprit* (mind or spirit) is tighter in French than in English. English does have a sense of salt as "sharp humor or wit," as well as the expressions "Attic salt" or "Attic wit," which also occur in French. But a primary figurative meaning of *sel* is that which gives flavor to works of the intellect, a liveliness of mind that spices up or refines a discourse or thought.—Trans.]

Even a tourist out on a *mehari* (a trek on camelback) quickly learns a survival technique: his eye is trained along the way, educated to a diversity that may surprise those who mistakenly believe a desert expanse to be monotonous. It is a genuine museum of minerals, with all sorts of nuances in color and unexpected textures.

A first surprise: the landscape changes constantly. The eye quickly learns to pinpoint all that hails it in the vast expanse in which it moves: the colors of the stones, the whitish efflorescence of a salt leaving, another bright yellow one that recalls crude sulphur, the gleam of iron, the entire palette of mineral oxides, but also the indefinitely varied forms of the dunes, reminiscent of a romantic composer's "Theme and Variations," and a whole profuse herbarium of blades of grass and tiny flowers, to say nothing of the climbs and descents all along the way.

The desert resembles a series of pages covered in signs, which the Tuaregs read with the professional ease of an art historian flushing out the many symbolic meanings in a painting. A whole scattered script is ranged about with the utmost clarity, presenting itself for the mind's exercise. The desert is clean; all its fauna, animals great and small, are scoured endlessly in a dry cleaning yet made more perfect by the bleaching sun. Animal skeletons, shells of automobiles, and abandoned trucks are like sculptures along the trail, marking it out even more.

Contemplating this cleanness for hours and days on end, the eye-mind fills with it bit by bit. Lucidity attains the eye-mind. All becomes sharp, clear, incised. Thoughts gain so much precision that they ossify and then mineralize. They crumble into grains of thought, into indivisible units that the mind turns on their various sides, admiring their hardness, texture, and reflections.

Mind of salt . . .

SAINT-JOHN PERSE.   This great French poet of the twentieth century, a winner of the Nobel Prize in literature, was obsessed with images of a severe, parched, dry earth sparkling with minerals. The works of Saint-John Perse abound in references to salt, a compound from the intersection of the mineral and the organic. Its very existence evokes truly ancient times, when (according to a scenario such as Oparine's) life was seeded in a pool of brine out of a primeval soup: "Everything is salty, everything is viscous and heavy like the life of plasmas."[9] It is also a reminder of tears from our infant days: "I remember the salt, I remember the salt my yellow nurse had to wipe away at the corner of my eyes."[10] Likewise, salt also recalls the salty nature of secretions related to sexual pleasure—"Power, you sang on our roads of splendour. In the delight of salt the mind shakes its tumult of spears. With salt shall I revive the dead mouths of desire!"[11]—between impregnation and mortification, as some age-old books instruct us: "So goes all flesh to the hairshirt of salt, ashen fruit of our vigils, dwarf rose of your sands, and the spouse of a night shown out before the dawn."[12] With this, Saint-John Perse makes himself the poet of the tribe, the awakener of ancient archetypes: salt is linked to the caravan, to the desert, and to thirst. Here, the instances of assonance in the word pairs hair shirt–salt, salt-sand, and desert-desire [The French word pairs are: *cilice-sel*, *sel-sable*, and *désert-désir*.—Trans.] are seeds of the poetic:

> In the delight of salt the mind shakes its tumult of spears. . . .
> With salt shall I revive the dead mouths of desire!

> Him who has not praised thirst and drank the water of the sands
> from a sallet

> I trust him little in the commerce of the soul. (And the Sun is
> unmentioned but his power is amongst us.)[13]

The second verse might also have been written by Saint-Exupéry, whose *Terre des hommes* and *Citadelle* time and again return to the Stoic themes of detestable satiety and blessed thirst.[14]

> Thus salt calls the exile to the outer reaches:
> You traffic not in a salt more strong than this, when at morning
> of kingdoms and omen of dead waters swung high over the
> smokes of the world, the drums of exile waken on the marches

> Eternity yawning on the sands.[15]

A biographical factor is pertinent here: Saint-John Perse's character was fortified by a long stay in China from 1916 to 1921. He took advantage of this time to make long treks, even into Mongolia, and he read closely some of the best of that period's sinologues and specialists on Tibet. This fact encourages one to read Saint-John Perse in alternation with pages—also inspired by the Asian deserts—drawn from the works of Victor Segalen or Paul Claudel.

Though the poet most often identifies salt with its ubiquity—owing to the physiological need for it and its effects on organisms—and with the rhythmic regularities of its combined ingestion and excretion, he also on occasion limits himself to its mere mineral presence, if a curious one still enlivened by a spectacular chemical nature that brings about considerable releases of heat in experiment: "Wisdom in the foam, O plagues of the mind in the crepitation of salt and the milk of quicklime!"[16]

For salt is valued as much for the purity of its crystal, "the idea pure as salt,"[17] as for being identical with its flavor, for its primal simplicity. In a 1960 study, Jean-Pierre Richard relates the meaning of salt for Saint-John Perse to the sensations of dryness, thinness, and sharpness, even to the attractive draw of nothingness.[18] These were all familiar feelings to the first settlers in the United States.

WEST SALT STORY, 1650–1850.

Pioneers came first,
looked around:
the animate landscape,
the open sky,
friendly tribes, undulating
expanses of land, tall trees,
big rivers.
Government-mandated surveyors.
They looked, they mapped out,
they charted and inscribed: place names,
brooks, hills, springs, lived-in
settlements, woods, lodes,
deposits, lignite, iron, sulfur,
salt licks, Indian trails.

Then came the industrious
Germans. And the French,
nimble trappers,
made themselves scarce.
Settlers logged. Wood
was turned into barrels,
to hold bushels of salt
boiled from brine:
more trees fed the fires.
Population came in droves,
flocked to the frontier,
to Ohio and to Kentucky,
onward to Illinois and Michigan,
and yet more salt,
more and more salt,
was needed.
Packs of mules followed,
deepened Indian trails
into salt routes. Meanwhile,
to keep the fires going under
the huge kettles, coal was
dug out and transported
from afar.
Boiling salt followed
British tradition. More and more
people arrived.
Wilderness receded, a
horizon at a time, towns
incorporated and named
themselves. The more daring dug
for salt. Mining was tried,
and they struck salt.

SALT ROUTES.   France offers a gamut of examples. As throughout the Sahara, salt routes sometimes hundreds of kilometers long link centers for production of the precious commodity with its consumption out-lets. For this reason, the hinterland serviced by the Mediterranean salt-works—which include the Berre area, the Camargue, and the saltworks of Hyères, at least during the five or six centuries for which there is documented evidence aplenty—extends northwest to the Cévennes up to the Central Massif, north as far as Queyras and the Barcelonnette and Briançon regions, and east and northeast as far as Turin.

The seasonal movement of sheep flocks, accompanied by mules or donkeys to carry provisions, along tracks traced between highlands and lowlands from time immemorial, gives us a sense of this movement of salt. These trails include delicate, acrobatic passageways; some have survived in their original condition, as today's hiking paths, without having been converted into roads (or highways). The mules, with their sometimes bulky loads, made their way along the very edge of the abyss so as not to risk brushing the rock face.[19] One example of such a trail is the path going from Seyne-les-Alpes to Barcelonnette by way of the Bernardez Pass, impassable in winter.

For centuries on end, the routes remained unchanged. The saltworks of Peccaïs, near Aigues-Mortes, and other saltworks of Languedoc supplied the Rouergue.[20] The saltworks on the Berre lakeshore served Aix and the Durance river valley (Manosque, Valensole, and Sisteron, and then Seyne, Saint-Vincent, Barcelonnette, Gap, and Briançon).[21]

This circulation of salt and other goods took place very early on, during the first centuries of our era, using the Roman road network to complement transport by sea. It answered the need to transport salt produced in the saltworks of the Toulon and Hyères regions from Nice to the Piedmont region after its shipment to Nice by boat.

Thus Emperor Augustus, after having subjugated the Ligurians, had built a road following a northerly course along the Roya river valley from Ventimiglia to Borgo San Dalmazzo, by way of the Tende Pass, following sheep tracks used for transhumance between coastal Liguria and the Roya's high mountain valley. This and many other routes vanished from the map from approximately the ninth to the thirteenth centuries: the invasions of the Goths, Lombards, and Saracens forced communities to leave the valleys and settle in the heights and to block access so as to protect themselves better from invaders.

At the start of the thirteenth century, with civic peace restored, the counts of Ventimiglia rebuilt the road linking Ventimiglia to the Piedmont region via the Tende Pass, a road that, together with its revenues, fell into the coffers of the counts of Provence at the close of the thirteenth century. But in 1388 Amedeo VII, count of Savoy, seized the Piedmont region and the Roya valley, acquiring a monopoly on all the salt that passed through the valley. This was the case until 1407, when the right-of-way fees demanded by the count of Tende were prohibitive enough to compel the lords of Savoy to seek an alternative way to move salt from Provence to the Piedmont region. They thus built another road, further west, going from Nice to Turin by way of the Vésubie river valley.[22] The drawback of this new *camin salié* (literally, "salty trail," in Provençal) was its seasonal passability, since the Fenestre Pass is closed during the winter, while trade in salt takes place above all in winter.[23] The main road continued to follow the Roya river valley during the fifteenth and sixteenth centuries, while a secondary road was constructed along the coast between Breil and Menton to link other unloading points for Provençal salt. But the local lords continued to collect right-of-way fees, which disrupted trade. Among other things, this prompted an order to the seneschal of Provence from Charles VIII, king of France, prohibiting the count of Tende from imposing an exorbitant toll (March 17, 1491).

In 1581 the house of Savoy annexed the county of Tende. Almost immediately (1592), the dukes of Savoy decided to build a new road to link Nice and the Piedmont region, bypassing the difficult Brouis Pass and the villages of Breil, Saorge, and La Brigue. Construction was completed in 1643. A new road passable by coach and going from Nice to Cuneo through the Roya valley was built at the end of the eighteenth century by Duke Victor Amedeo III. In 1892 it took ten and a half hours by coach to cover the eighty-two kilometers from Nice to Tende, and the trip cost seven francs or the equivalent of one hundred fifty francs in 2001.[24] Further on, the road went by way of Limone via the Tende Pass Tunnel, completed in 1883, to end fifty kilometers away in Cuneo.

In this way, one can trace the gradual transformation, from the thirteenth to the twentieth century, of a mule drivers' trail into a coach road and then into a railway, with the concomitant increase in the traffic of salt and goods. In 1776, 18,317 mules allocated to general commerce left from Nice; 16,124 were bound for Cuneo and 2,178 for Turin. In the same period, 30,000 to 35,000 mules loaded with salt took the same route. Salt production levels were relatively constant from the fourteenth to the nineteenth century: according to a 1334 survey, yearly production at the Hyères saltworks was between 8,000 and 11,000 metric tons;[25] in 1892 it was still 10,000 metric tons.[26]

Such trade was of considerable importance and posed a number of problems: in addition to those concerning road passability, security, and the levied tolls were the problems of the availability of pack animals and the freightless return trip. On this last point, convoys returned carrying wood, manufactured goods (cloth, fabrics, metal objects, ceramics, glassware), and foodstuffs (spices, salt meat or fish, fruit, etc.). Most frequently, the pack animals were requisitioned in villages along the way.[27] Modern states were born of these mule convoys, as were certain modern lines of communication (the highway from Marseilles to Sisteron follows the old salt road): ensuring the safety of men, pack animals, and goods, as well as the security of the salt depots—and granting oneself the genuine authority to requisition—demanded strong political power.

The salt routes thus allowed for economic exchange between regions, wood from the Central Massif and the Rouergue region being sent down to shipyards in the Languedoc in exchange for the supply of salt to Rouergue salt depots. They allowed the establishment of economic powers, such as Venice on the Adriatic or the Hanseatic cities in northern Germany. In that area, Lüneburg salt was transported to Lübeck, where it then was exported to Scandinavia or shipped off to the Shetland Isles, north of Scotland, there to be used in curing herring; the boats of the merchants from Hamburg, Kiel, and Lübeck returned loaded with fish. This salt route also promoted cultural exchanges: Johann Sebastien Bach, who studied in Lüneburg from 1700 to 1702, availed himself of it in 1705 to go hear Dietrich Buxtehude play the Marienkirche organ in Lübeck.

The need for salt is ubiquitous and constant. It is only quite recently, in the twentieth century, that it has become mundane. Prior to this, it was rare, a precious commodity; not for nothing was it referred to as "white gold." Hence a Provençal saying from the start of the sixteenth century held "Cant an de sal non an dol" (When they have salt, they don't have oil),[28] making allusion to the uncertain supply of two ingredients essential for Provençal cookery: salt and olive oil.

In fact, sayings and proverbs, condensed in order more easily to be committed to memory, conceal an entire body of knowledge that must be extracted so as to restore their original breadth. This is the case with the following proverb.

THE PROVERB OF THE TARDY SALT.   The Dutch proverb "Hij komt met het zout als het ei op is" finds its French equivalent in the proverb of the *carabinieri* (Italian military police) who always arrive too late. [This refers to the French expression, "arriver comme les carabiniers," meaning "to arrive too late."—Trans.] The literal translation is: "He brings the salt once the egg is eaten." The saying gives existential significance to a mundane feature of daily life. It deals actually with the importance of offering help, or at the very least counsel, at the right moment, when it is needed, before it is too late.

Salt, in proverbs often associated with eggs, functions here as an indicator of sociality.[29] As in the words of Jesus (the salt of the earth), salt, which imparts flavor to all foods, takes on the metaphorical meaning of all that gives value to relations among human beings. It is written in the Gospel of Matthew (5:13): "You are the salt of the earth; but if salt has lost its taste, how shall its saltness be restored? It is no longer good for anything except to be thrown out and trodden under foot by men." This passage comes right after the Sermon on the Mount.

Note the structural analogy between the Dutch proverb and Christ's pronouncement: each emphasizes a total devaluing of salt. In both cases, salt symbolizes a bond between human beings, their brotherhood, and above all a moral conscience, the conscience to bring aid to the other, to help the other when in need. And what about our brotherhood of the biosphere, with fellow creatures?

ALPINE SALT.   Humans and animals alike need salt. To return to the contrast between settled peoples and nomads, if agriculture determined the settling of human populations, then the breeding of livestock has had continued importance for both preserving a version of nomadism and linking the two ways of life.

Grazing animals need to be supplied with salt no matter where they are. During transhumance, salt is perhaps the essential provision. From my childhood in Grenoble, I recall the unending flocks of sheep that unfurled through the city on the way to their lofty destinations, each trailed by a shepherd with a small donkey cart carrying essentials, most crucially blocks of salt (today this is done by truck).

Recently, I observed the behavior of a few cows at pasture on the Aubrac Plateau, at an elevation of 1,200 meters. Let me paint the picture for you.

These domesticated animals are like prisoners. Members of the herd, they know each other all too well. Some ignore one another carefully; others are friends. Incarcerated livestock, to be sure, but in what a landscape!

It is an immense meadow. A path cuts across it, paralleling the watershed. Below is a cluster of trees, some already nipped by a late freeze; their leaves, mingled at this close of spring with those of the unaffected trees, are already autumnal. Bells chime. The cows graze. They stray quite far, in constant motion, meandering here and there.

The paths that the cows take about the meadow are not completely random, however, even though there is plenty of grass (it has recently rained). The cows coordinate their moves. Some animals dominate others. The older ones exact respect from the younger ones.

One lively, even impetuous, heifer has already made a number of attempts to lick the salt block set in the middle of the meadow. She was knocked out of the way several times by a cow that took her place or took it again. She now stands at a distance, a distance calculated so as neither to offend nor to lose her her spot once she finds herself at last alone. The salt she covets also represents her freedom to act, a suspended bit of time. She makes penance before at last allowing herself, almost surreptitiously, her little treat.

LICK. American English has a noun *lick*. According to the *Oxford English Dictionary*, among other meanings, it is "a place to which animals go to lick earth impregnated with salt." Indeed, Christopher Gist wrote in 1750–1751 in his diary of "several Salt Licks, or Ponds." He was a professional surveyor for the Ohio Company of Virginia, and he was reporting on an area west of modern Zanesville, Ohio. He also made note of salt licks in Kentucky. Lewis Evans, at around the same time (1755), brought out a map of the middle British colonies in America, based on information from fur traders. It shows, also in Kentucky, a Great Salt Lick River with, as a place name that gave rise to the modern Licking, the Great Buffalo Lick.

During the Indian Wars on the then frontier settlements, Daniel Boone was captured. His captors brought him to Jackson Licks, Ohio (it was then named Salt Lick Town). Even though Indians knew very well how to use salt springs to provide them with salt, they were testing the white man's knowledge, his survival skills in the wilderness.

France, too, preserves on its map a past concern with the availability and production of salt.

LIKE THE DAWN.   Many places have long been dedicated to salt, as the vocabulary of place names reveals. Prior to presenting this geographic vocabulary, a minor point of information must be mentioned or recalled: a common linguistic transformation is the change of the vocalization of the phoneme written -al into the phoneme -au (seemingly by way of a phoneme *ao*).[30] A prototypical example is the Latin adjective *alba*, white, which in French yields the word *aube*, or dawn, as well as common nouns like *aubépine* (hawthorn) and *aubier* (sapwood) and place names such as Aubagne (white bath), Aubespeyre (white stone), Aubenas (white nose?), Aubin, Aubisque (white peak), and Aubusson (white wood).

Other examples are the many words with the prefix *mal-* (ill- or mal-), having, as it were, *mau*tourné (taken a turn for ill), from *mauvais* (bad) to *maugréer* (to grumble), by way of *maudire* (to curse), and the rugby *maul* from the English "to maul," *malmener* (to ill-treat), or even *maussade* (disagreeable) (in Old French, *sade* meant "*agréable*," or pleasant).

Likewise, the prefix *val-* (vale) is found in Vauvert, in the expression *à vau-l'eau* (literally, "a downstream water flow," or "with the current"), and in Vaudois (Valdo), Vaucouleurs, Vaucresson, or Vaux-le-Vicomte.

Since salt in Latin is *sal*, it is not surprising that a certain number of compound words should likewise have acquired a *sau-* syllable, such as *saumure* (brine) and *sauce* (sauce, "salted" *salsus*), *saucisse* (sausage, "salted" *salsicius*), *saunier* (saltworker), *saupiquet*, a spicy sauce or stew (from *sel*, salt, and *piquer*, to be spicy), *saupoudrer* (which literally means "to powder with salt" [Today, the meaning of *saupoudrer* is simply "to sprinkle with" or "to dust with" and is not restricted to salt.—Trans.]). Among place names are the very clear examples of Saulieu and Lons-le-Saunier, a city of mineral springs whose tourist brochures extol the virtues of salt water. In the technical lexicon of the Guérande saltworkers, *saumater* means "to become salty."[31]

TECHNICAL VOCABULARIES.   I will now lay out the technical vocabularies used in the chief centers of salt production in France. Study of the lexical traces of these different technologies shows them to be autonomous and independent of one another, as if people had had to reinvent the wheel at every juncture.

To start, I will examine the lexicons peculiar to the salt marshes of the south of France, to the Guérande salt flats in southern Brittany, and to the igneous saltworks in the Franche-Comté (eastern France). My aim is to furnish a representative, rather than exhaustive, depiction and to elucidate the origin of some of the terms employed. Then, following this, I will sketch out a comparative analysis of these three sets of terms.

The Salt Marshes of Southern France. The *aiguilles* (literally, "needles") are the channels ensuring water circulation among the various evaporation compartments, or crystallizers, where the salt crystallizes.[32] The origin of the term is unknown. One could perhaps associate it with a term from Venetian dialect, *agger*, used in the Middle Ages to designate the dikes around the perimeter of the laguna at the Chioggia salt marshes.

The *jas*, from the Old Provençal *jatz*, derived from the vulgar Late Latin *jacium*, or "site," is the first basin, used for decantation of the supply water that deposits its mud and debris there.

The word *partènement* is important. It designates the crystallizers. The *Trésor de la langue française* (Treasures of the French language) reports its origin as unknown. I would relate it to the Latin *partitio*. The construction of the word *partènement*, parallel to the construction of *soutènement* from *soutenir*—or even parallel to the construction of *tenir*, related to the English *tenement*, or parallel to the pair *appartenir-appartement*, or even to the English variation of *contenir*, namely, containment—suggests a vanished verb, namely, *partenir**, meaning "to keep apart," which matches precisely the function of the *partènement*.

Cairel is the name for the earthen levee that marked out the crystallizers. Perhaps the word is related to the word *carré* (square), an allusion to the rectangular contour of the crystallizers?

Ardes are small earthen levees, certainly so named for their steep slope. [*Ardu* is an adjective meaning "steep" or "difficult to climb."—Trans.]

The *martellière*, whose name comes from *martel*, Old French for "hammer," is a sluice gate allowing control of the water circulation among the various basins.

The heaps of salt are called *javelle* (swath) and *gerbe* (sheaf), in chronological order and increasing order of size. The first of these terms is of Gaulish origin, meaning that which is gathered by pile or handfuls. According to the *Trésor de la langue française*, a *javelle* is "the armful of grain . . . reaped with a scythe or reaper and kept in small piles on the stubble fields before being sheaved." Clearly, the terms used by the saltworkers of southern France are borrowed from agriculture: *javeler* (to make into swaths) refers to the raking of the crystallizers to gather the salt.

In the same way that various agricultural activities are called *abattage* (slaughter), *moissonnage* (harvesting), *battage* (threshing), *épandage* (spreading), *émondage* (pruning), formed with the common suffix *-age*, so the saltworkers of southern France practiced *allégeage* (partial or complete draining of water in the crystallizers), *battage* (breaking up the layer of salt and making piles), *coupage* (as with a wine, diluting a concentrated brine), *démarmaillage* (breaking up the crust of salt with a pick), and even *levage* (transporting the swaths or sheaves to secondary salt piles).

Finally, an *échelle*, or ladder, helped in dumping out the salt on the enormous central pyramid-shaped pile called either a *camelle* or an *haricot*, depending on its shape. The *camelle*, from the Provençal *camello*, or camel, owes its name to the irregular profile of its ridgetop, which was sculpted by the wind and rain. The *haricot* (bean), whose contour at its base is that of a bean, comes from the verb *harigoter*, which in Old French meant "to tear up, to tear to shreds."

Guérande (Artisanal Salt Marshes of the Western Atlantic Coast). At Guérande, the supply channel for the saltworks was called an *étier*, related to the Latin *aestuarium*, "estuary," meaning a large flow of water.[33]

The first compartment, used for decantation, is called a *vasière*, whose meaning, I dare say, is obvious. [*Vase* is French for "mud, mire."—Trans.]

The basin for the first stage of condensation includes several compartments and is called the *cobier*. Though one cannot do so with absolute certainty, linking this name—as do Jean-Claude and Jacqueline Hocquet—with the Latin *corbulum*, a term with an analogous meaning in the Venetian dialect of the medieval Chioggia saltworks, seems to me an elegant supposition.[34]

The *cuy* (or *cui*, or *coëf*) is a channel fitted with a sluice gate that allows one to let water into or out of a compartment. The word is of Celtic origin.

The salt marsh has other channels, called *bondres* (or *boudreaux*), that are offshoots of the supply channel bringing seawater into the salt-works. Suggested similar etymologies would link these terms either with the etymology of the word *bonde* (a tap hole in a cask, barrel, or keg, for example) or of *fondrière*, a pothole.

The word *fare* indicates the reservoir around the perimeter of a saltworks; its origin is unknown.

The *oeillet* is the compartment designated for crystallization. It can be related to the Old French verb *ouiller*, of which we have a confirmed usage from 1322. This verb is a contraction of *aouiller*, literally, "to fill up to the eyes"; it means to resupply the liquid in a cask as evaporation takes place.

The *ladure*, a circular platform formed in the center of the *oeillet* by an expansion of the *barrure* (a kind of lock), receives the daily salt yield. Evidently, it corresponds to the term *piadura*, used in the medieval Chioggia saltworks.

Saltworks in Franche-Comté. The metal cooker in which brine evaporation by means of heating takes place is called a *poêle* (pan).[35]

Each *poêle*, accompanied by a *poêlon* (skillet), occupies a separate space, called a *berne*. This term is derived from a Germanic word, *berm* in Dutch, *bryne* in Old English, and *brine* in contemporary English.

A *remandure* is a series of sixteen consecutive *cuites* (boilings) over the course of a dozen days and nights of constant supervision of a single pan. (The *remandure* has the sense of the English *remainder*, or *leftover*, that is, that which is permanently left over.) The *cuite* is the cooking period.

Cooking time is divided into four stages. The first one is the *ébergé-muire* (which one may be tempted to analyze as a "*héberger-muire*," *muire* deriving from the Latin *muira*, *saumure*, or brine). [In general usage, *héberger* means "to lodge, house, shelter, or accommodate."— Trans.] After this come the *premières heures* (the first hours) and then the *secondes heures* (the second hours). The final stage is the *mettre-prou*, the period necessary for gradual crystallization over low heat. I understand *prou* as in the expression "*peu ou prou*," where it means "much" (it shares a common root with the English adjective *proud*, from the

Late Latin word *prode*, meaning "advantageous"). The *mettre-prou* is the phase of intense salt crystallization.

Following this, some female saltworkers collect the salt. Others clean the pan, which is called *râbler* (chipping, or "dodging"), a term derived from the Latin *rutabulum*, the verb *ruere* meaning "to precipitate, to make fall." This *râblage* (cleaning) removes the *schlot*, from the German *Schlotte*, that is, the encrusted deposit, or scale, left on the bottom of the pan after the cooking has taken place.

These varying sets of terms are striking in their richness, in the beauty of all these technical terms, and in the regional particularities that they express. As one might expect, the lexicon of the saltworks of the south of France abounds in words derived from Provençal, while that of the Guérande marshes is characterized by the Celtic origin of many terms. At Salins, on the other hand, one finds, as expected, terms with German resonances such as *berne* or *schlot*. Conversely, the fact that certain terms from the marshes of the Western Atlantic (*cobier, ladure*, and even *vette* and *morts*) can be aligned with those from the medieval Adriatic saltworks suggests that technological expertise was imported during the initial development period of these Atlantic coast salt marshes prior to the first millennium.

One is impressed by the differences in the technical lexicons from these three salt production areas. I would describe them without hesitation as three regional dialects spoken by workers responsible for salt production.

Consider in particular the term that names these workers in each of the three regions. It will supply the key to the linguistic system governing each of these lexicons. In the saltworks of southern France, the worker is a *saunier*: the emphasis is on the harvesting of salt. In the western saltworks, in Guérande, among other places, one speaks of *paludiers* [*Palud* or *palude* is a noun meaning "marsh."—Trans.]: their expertise pertains to the marsh, that is, to the management of either the circulation or stagnation of water. Finally, the igneous salt of Franche-Comté, in particular that from the saltworks of Arc-et-Senans built by Claude-Nicholas Ledoux, is produced by the *berniers*, that is, by artisans who specialize in evaporating brine in heated cauldrons.

This very simple observation provides a vital clue. In the saltworks of southern France, it was a question of harvesting the salt; hence the terms common to agricultural labor—*javelle, gerbe, camelle, haricot*—

and all those verbs that produced the substantives ending in -*age*: *battre*, *démarmailler*, *lever*, *alléger*, and so on.

In the Guérande saltworks, making water circulate is expressed with a great many terms: in Breton, *water* is *dour* (or *deur*), and *dourer* means "to supply an *oeillet* (crystallization compartment) with water." The *dourure* is the amount of water introduced into a crystallization compartment in the supply process. The supply channels for the successive basins are the *étiers*, *bondres*, and *cuys*. And there is also the *dlivre*, a gutter running parallel to a row of crystallization compartments (the likely origin of the term is the verb *délivrer*, "to deliver").

And water is contained in these places, which themselves are diverse and represented with great lexical variety: *vasière*, *cobier*, *fare*, *oeillet*. In Breton again, *darn* is a partition; *aderne* (with the privative "a") is the final evaporation compartment prior to the crystallization chamber.

In the saltworks of Franche-Comté, it is a matter of cooking a brine. Workers are called *berniers*, as in the corresponding English term, reported in the seventeenth century, *briner*, that is, salt boiler, saltworker.[36] Nor is it surprising that the description of the work is akin to that of chefs and scullions in a kitchen. A division of labor prevails there: besides the *berniers* (salt boilers), female saltworkers, the *femmes de berne*, remove the coals from the fire; they are the *tirari de feu*. Other female saltworkers extinguish these coals with water; they are the *etaignari*. Both sorts of female laborers draw the salt out of the pan once it is formed; they are thus called the *tirari de sel*. Four women shape and dry the cakes of salt, called *salignons*: the *mettari* fills the mold; the *fassari* fashions the salt into its shape; finally, the two *séchari* dry the salt cakes over the coals before a male saltworker comes to *enbenater* them, to load them in a *baneton*, or basket.

Three centers of salt production: the Mediterranean, the Western Atlantic, and Franche-Comté. Three specific lexicons, showing little or no overlap. Three dominant metaphors: the harvesting of grain, the management of marshland, and the cooking of brine, which is supplied by one or more sources of salt water. Three visions, irreducible to each other, of what is but one physical process of evaporation-crystallization. Can one go so far as to say that these are three groups of human beings whose visions of the world in all likelihood differ in a great many other respects? Is translation between these various technical languages, each of which forms a coherent whole, possible?

THE PROVERB OF THE BLAND EGG. Let me end this chapter with another saying, about blandness. A Portuguese proverb runs, "Ovo sém sal, não faz bém no mal" (The literal translation is "an egg without salt does neither good nor ill"). As with many other proverbs, the assonances help in committing this one to memory: *sém-bém* and *sal-mal*.

Removed from its context, this proverb has an enigmatic quality: that an egg eaten without salt does neither good nor ill, that it has a neutral effect—if this ancestral folk wisdom is to be believed—leaves one cold or nonplussed.

But the virtue of a proverb lies in its implicit meaning as well as in the metaphor it explicitly expresses. Here, the egg is considered more generally representative of food. An insipid dish can do no harm, whereas some food, from the mere fact of its saltiness, can be harmful. So, one way of understanding this proverb is as a ratification of a salt-free diet.

But the saying certainly has a much greater significance, since salt in a more general way symbolizes that which imparts flavor or spice to existence. One must thus understand it above all in a moral sense: to live without emotion or passion is to live without risk, day to day, deriving neither satisfaction nor sorrow from living.

This broader interpretation is bolstered by the fact that the egg (think of Easter eggs) symbolizes fertility and the living being. A life without passion is certainly dead boring!

But giving flavor to life, starting with dishes at the table, requires at least a pinch of salt, which in history did not always come cheap or easy. What about geography and the geopolitics of salt? How and where was the grain of life harvested? How did mankind go about procuring the precious commodity? This is the theme of the next chapter.

three **harvesting**

Salt crystallizes by evaporation from salt water, a process that can be shown to children using a saucer. The same process affects lakes (the Great Salt Lake) and even seas (the Dead Sea, the Aral Sea). Over the course of geological eras, the deposits of evaporated salt, or evaporites, are buried beneath sediments, but a fluidity unexpected in such a crystalline solid, related to its lower density, causes an upwelling of salt through fissures in the earth's crust that forms immense mushrooms whose caps, near the surface, are worked as rock salt mines.

    But is such a mining industry a genuine source of wealth? Montesquieu poses this question regarding the Spanish bankruptcy (to which I also allude in chapter 4), a surprising event, given that Philip II's kingdom seemed to have grown rich on all that gold from the Americas. This is still an issue today. Mining wealth remains illusory; perhaps, human labor is the only real wealth. This chapter's reflection on economics will be completed by a Chinese proverb on salt, dating from the Ch'ing dynasty, the deep meaning of which concerns the virtue of effort.

And indeed the labor of the saltworker, that artisan who extracts sea salt in salt marshes, is painful, strenuous, and constantly threatened by the possibility of poor weather. Balzac describes this hard, unrelenting labor in the opening of his novel Béatrix. *One can read in it a metaphor for the work of the writer, facing his white page like the salt makers of the Guérande before their heaps of salt.*

Separating salt from the water in which it has dissolved also serves, conversely, to supply potable water. In California and Saudi Arabia, desalination of seawater has become a technology essential to the establishment of human communities in regions with desert climates. But how did deposits from which we mine the coveted mineral form?

SALT DOMES.   Did this occur during the Jurassic Period, 150 million years ago, or earlier, 250 million years ago, in the time of Pangea, our single continent? At some point, a sea withdrew, and after the sun had prevailed over all that water, salt was left behind. This evaporite then vanished from the earth's surface; the erosion of mountains covered it over with debris. The passage of time, the span of geologic time, sank the evaporite, and megatons of sediment came along to bury it, to make it into a salt deposit.[1] Here lies salt. But a stone that sleeps long comes back strong! Salt sometimes finds itself lighter than the gravestone that sits atop it, weighting it. So, it escapes. Like Houdini, salt, beneath all the latitudes, paradoxically endowed with an upward force by all that piling up of marl, sandstone, sand, and mud on top of it, strives to shoot up through all the faults and fissures. It weaves its way in, creeps through, and rises.

You rightly may be astonished to learn that salt is plastic: under pressure, it flows upward, like toothpaste in a squeezed tube. Such flow is called creep.[2] As a result, a whole forest of columns grows up vertically, made of gigantic fingers of salt, or diapirs, of very large underground temples, composed of pedestals that are crowned, like a column by a capital, with bulb-shaped domes.[3] These caps lend a sort of cryptogamous look to the diapirs in their underground world, like mushrooms in the underbrush. And likewise the diapirs, rising surreptitiously—one millimeter per year over millions of years—appear to grow and multiply.

In the Kavir Desert, in the very heart of today's Iran, twelve diapirs have managed to entwine and join their bulb-shaped domes, creating a splendid entablature forty kilometers in diameter.[4]

Though the salt columns rise, the vault that at first floats above them slowly collapses, and the whole structure takes on the appearance of cells in a bee hive.

How beautiful the earth is! The air traveler and the satellite's camera see these structures from above. They form regular flagstonelike patterns of frets and friezes, with occasional concentric striations as in some hard candy, or sticks of Brighton rock, or those very pretty multicolored candles made in Chester.

When the sediments covering them are not very thick, the diapirs pierce them and continue to rise. These ostentatious displays, eruptions that become visible when the salt, since there is such an abundance of it, outlives the rains, are called namakiers. Namakiers are for salt the equivalent of glaciers for water (*namak* means "glacier" in Farsi). The Mount Zagros namakiers in southern Iran form a decorative fringe and border: this is how the earth displays its visible history!

To feed itself, humankind draws on this dramatic mineral array; salt mines are nothing but drillings carried out for centuries in these salt domes. Isn't it an enviable resource, indeed, a source of great wealth, for a country to own one or more such underground beds of salt?

The abundance and density of the salt beds excite prompt hyperbole and elicit superlatives: I draw from a primary school science textbook, nearly one hundred twenty years old, its description of the Wieliczka mine—at a time when nearby Auschwitz was a village like any other—since this description is representative and, one could say, canonical:

> It is a string of enormous underground chambers, an immense town with its own streets and public squares. The huts for the miners, and stables for the necessary work horses, are carved out of salt. There is a large population there, and hundreds of workers are born there, and die there without ever having left their underground chambers, without ever having seen the sun's light. There are chapels for worship services and many of the galleries are loftier and broader than churches. A great number

of lamps are always kept burning there, and their flame, reflected in every direction upon the salt walls, makes the walls look at times clear and sparkling like crystal, and at other times shine with the most beautiful colors.[5]

Are such salt mines extremely profitable, then?

MINING. Beginning in about 1728, Montesquieu, who in 1748 would publish *L'Esprit des lois* (*The Spirit of Laws*), worked and reflected on his masterwork. He drew up a set of notes, which remained in manuscript form, entitled at first *De la principale cause de la décadence de l'Espagne*. In it, he studied the paradox that Spain, though rich with New World gold, was forced into bankruptcy during the reign of Philip II: "Spain derives little advantage from the great amount of gold and silver it receives every year from the Indies; at first, the gain was considerable, but it was destroyed by itself, and by the inherent flaw of the thing; I will explain my thinking," Montesquieu wrote, presenting the problem. He begins the account with a distinction between goods that have a "natural use," such as wheat, wine, or cloth, and other goods that have a "fictional use," since they are by nature monetary signs. Montesquieu announces the conclusion to his argument immediately following this distinction: "Having conquered Mexico and Peru, the Spanish abandoned sources of natural riches for fictional riches, and the fantasy of instant profit made of them complete fools."[6]

The steps of the argument are: (1) the gradual erosion of the value of American gold and silver as its volume increased with its arrival in Spain on galleons from America: this is a near-mechanical effect, with an overabundant supply of a commodity lowering its price; (2) economic war (as we would say today) with other powers, principally England and the Netherlands, whose reaction, in the face of the Spanish monopoly on American gold, was to create new means of payment: "public credit took the place, for them, of mines," in Montesquieu's fitting formulation; (3) the great distance between Spain and the mineral resources of Mexico and Peru, accounting for considerable mining and transport costs; (4) the axiom that every sovereign holds his wealth only "as a consequence of the affluence of his subjects"; (5) the fact that a state's gold and silver wealth has only symbolic value, sign value: paradoxically, the nations that possess the most gold or silver are also the weakest, the most vulnerable.

Let us note, before continuing, that historians have utterly confirmed Montesquieu's view. A study published ten years ago or so in the *Annales* demonstrated that the gold bars from America merely passed through Spain, spending only a few years there before ending up primarily in Dutch coffers!

But isn't Montesquieu's analysis now obsolete? Is it still true today that a country's mineral wealth is illusory? It would seem to be if we compare the standards of living in countries that lack mineral wealth (Japan is one of these) with those of countries with abundant underground reserves, whether these are Australia or Saudi Arabia. Was it true, throughout the centuries preceding modern times, that the possession of saltworks or salt marshes, that is, of exceptional mineral wealth, would have been a source of weakness, as Montesquieu claims? One can indeed accept this and can even hold that the trade in and taxation of salt were much more profitable than salt mining itself.

In short, only the labor of human beings creates lasting wealth. And this was also the case in the past: Cyprus's copper did not make it a great power, and Venice saw its period of greatest power after it had given up its own salt marshes, and after it had had diversified its trade to include products other than salt. Since I mention this example, let's not forget that Genoa, the great rival of Venice for centuries on end, never had its own salt marshes.

The popular wisdom of many cultures supports this claim; consider, for example, the following Chinese proverb.

THE PROVERB OF REJECTING THE BLAND. A Chinese proverb says: "Without salt, there's no end to blandness." Its meaning concerns the need to expend or invest energy (even just a little) if one wishes to resolve a problem. It dates from the beginnings of the Ch'ing dynasty, that is, from our seventeenth century.

A word, first of all, about the Chinese ideogram for salt: it has three parts. The entire bottom portion is a pictogram representing a salt marsh. The top left portion is the symbol for a piece of land, whereas the top right portion indicates—in a highly schematic way, compared to the ancient written forms—cubic crystals. On salt, this pictogram says it all!

But back to the proverb. It alludes to a moral concept, indeed, to a precept: do not dwell on appearances, which are insipid, stripped of meaning: "The salty and the sour each play a part in all that can be loved, but the supreme taste of salt—which is never-ending—is to be found inside."[7]

Like every proverb, this one harbors a powerful, even corrosive thought that time has so eroded and polished that only a truism has survived. The concept starts from the everyday observation that we hesitate when confronted with a food that is not salty enough or too salty (whence the saltcellar on the table, to remedy the first fault). Our perception of a lack of saltiness, which we experience as blandness, seeks correction through immediate compensatory action, almost instinctively.

The Chinese saying gives new vigor to the gesture of sprinkling salt on a dish to improve its taste. This common gesture appears insignificant; the proverb strips it of its ordinariness, compels us to consider it, rather, as the bearer of a much more profound meaning. It metaphorically turns this reflex act into a model for conducting one's life.

In this proverb, humankind is what imparts flavor to the world. In the end, each of our lives will be justified only by the strict standard of exactly what effort we have made to make our existence flavorful. The effort is its own justification, inasmuch as we make ourselves aware of it at the very moment in which we commit ourselves to it. And the fact that salt, this ordinary commodity, serves as this reminder of an ethics of frugality will not surprise us in Chinese thought.

Salt is not only extracted from mines, through the working of rock salt outcrops. It can also be harvested in salt marshes. And the Middle Kingdom already relied on these two resources well before the Christian era. This leads me to the rather outsized effort required for the upkeep and development of a salt marsh. I will describe this using French examples, some of which (the Guérande peninsula, near La Baule, or the salt marshes of the south of France) persist today.

SOLAR–EVAPORATION SALTWORKS. Agriculture takes advantage of photosynthesis to make plants grow or to maintain the grasslands on which to raise livestock. Saltworks are another use of solar energy, a soft energy and one that is free. This energy source permits—at least in favorable climates—other forms of energy such as wood or gasoline to be conserved; without it, each ton of salt produced in salt marshes would require the combustion of two tons of gasoline.[8]

Even so, the process requires the combination of a favorable geographic location (a coast with lagoons) and sufficient amounts of sunlight, with rainfall that is not too abundant: freshwater rains, which dilute the brine, are the haunting preoccupation of the artisans tending the saltworks. One parameter that determines the efficiency of a saltworks is thus the ratio of yearly water evaporation in millimeters to the total rainfall amount, expressed in the same unit measure. This ratio is approximately a factor of three for the Mediterranean saltworks (Hyères, Camargue). It is even more favorable in Australia (a minor salt producer, however, since it has a relatively small population, and exports are hindered by its geographic isolation), where it reaches a factor of twelve. Also in Australia, in the desert regions that receive only twenty-five millimeters of rain annually, evaporation is extremely quick: it takes only five days, whereas in regions receiving greater rainfall (two hundred millimeters), it takes an average of forty.[9]

Seawater is an aqueous solution of various salts, in which sodium chloride predominates. The average salinity of the oceans is about 3.5 percent (thirty-five grams of salt per liter of water), and one hundred grams of dissolved salt contain about seventy-seven grams of sodium chloride, ten grams of magnesium chloride, six grams of magnesium sulfate, and various other chemicals. Therefore, obtaining salt that is sufficiently pure is a multiphase process of crystallization: the whole science of the saltworker consists in evaporating the brine in the sun in such a way as to obtain successive separate crystallizations of the main dissolved salts.

These constraints dictate the layout of a saltworks. A supply channel of seawater feeds the basins, which are shallow but very large so as to maximize the surface for evaporation. The basins are linked and set out in a linear order, with a gradual increase in their brines' dissolved salt concentrations: the salt marsh employs a compartmentalization technique in which location specifies concentration (or, in like manner, density). The layout of canals and sluices permits the washing of the salt from the first crystallization with a saturated brine, so as to purify it without redissolving it. Saltworkers walk on the dikes around the evaporation basins to harvest these salts.

In other words, a saltworks embodies a dynamic of stagnation and flow. Take the example of the Guérande saltworks, eighty kilometers northwest of Nantes, that today still produces a renowned salt.[10] Water from the Atlantic arrives by way of a channel, eighty meters wide at its mouth and several kilometers long; eight main channels then branch off into canals called *bondres* or *bondreaux*. Water flow into the channels is renewed every six hours, with the tide.

The first basin, fed about every two weeks by means of a gate, is the sludge basin, with a surface area on the order of one hectare (2.471 acres). The water stagnates there for about fifteen days and settles out, at a depth of about thirty to ninety centimeters: as the basin's name indicates, mud is deposited there, in addition to other solid bodies, dead fish, and various detritus. The sludge basin is shaped like an overturned dinner plate, with a trench around its circumference to collect the sludge and all the debris, which the saltworkers remove from the sides.

From there, the water follows a winding, even labyrinthine course, supplying other basins called *cobiers* and *fares*, which are divided into compartments by small dikes. The water's depth is no more than ten or so centimeters in a *cobier* and about three centimeters in the *fares*. One of the reasons for these very shallow depths is that they allow ultraviolet solar radiation to sterilize and thereby clarify the brine. Biological activity is then limited to phytoplankton and extreme halophiles (see p. 100) in the microbial sheet coating the bottom of the *fares*. Water circulation occurs because the evaporation basins are arranged in succession on a very slight incline (on the order of one to five thousand at the most). The water takes three days to travel through all the *fares*, a distance of about a half-kilometer.

When the water is no more than about 28 percent of its initial volume, that is, between the third and the sixth basin, crystallizations of iron oxides, calcium carbonate (calcite), and calcium sulphate (gypsum) take place.

Last comes the harvest basin or *œillet*, where the water depth is no more than one centimeter. At Guérande, this basin yields the salt with the finest crystallization (the crystallization from the brine's surface, where vapor pressure is great): *fleur de sel*, that is, premium salt, cherished by connoisseurs. Of all salt production at Guérande, about one hundred tons of such premium salt are harvested, in comparison to ten thousand tons of the more ordinary gray salt. At first, the premium salt is red in color, from contamination by *Dunaniella salina*, a microorganism that thrives in this hypersaline environment and produces in great amounts the pigment beta-carotene, which also gives tomatoes and carrots their color. A saltworker harvests about three tons of salt a day.

The color of salt is worth commenting upon. A single crystal of pure salt is totally colorless. Impurities absorbed in salt crystals, starting with humidity, are responsible for its white color, through the diffusion, refraction, and multiple reflections of light inside the crystals. Salt harvested in the *œillets* is grayish because it crystallizes in vertical columns from the muddy bottom and is thus contaminated with particles of various minerals. Moreover, crystallization begins on the edges of the *œillet* and produces fine salt there. This fine salt has a greater surface area per kilogram than coarse salt and thus retains more impurities. Moreover, washing it is less effective than washing coarse salt, so there are more impurities—often trace elements precious for nutrition—in fine salt than in coarse salt.

I have referred to harvesting salt. In many respects, the work of the saltworker in fact resembles that of the farmer: in the diversity of tasks involved, the upkeep of the basins and dikes, the importance of weather conditions (having the foresight to let in enough water if you anticipate increased evaporation in the *fares* three days later; economic disaster in the event of prolonged rains), and so forth. Weather hazards can be catastrophic.

The Guérande saltworks carry on a regional tradition: for many centuries, the Bay of Bourgneuf, southwest of Nantes, was the site of intensive salt production. This salt was sought by Dutch and Hanseatic merchants, who used it to supply all of northern Europe (the British Isles, the Netherlands, northern Germany, and the Scandinavian and Baltic countries). In fact, the English language has retained the expression "bay-salt," particularly in cookbooks; it means "salt from the bay," with "of Bourgneuf" implied.

Regarding the Bay of Bourgneuf, note that it is the place of origin of the Acadians of New France: Cardinal Richelieu, who created the New France Company, had been abbey of the Assomption monastery, located on the Bay of Bourgneuf. He urged the inhabitants of the region to colonize Acadia, or New Scotland, today a region in Canada. Their familiarity with a marshy region proved invaluable to them: they built systems to drain the salt marshes using dikes and channels in order to prevent the large marshes from flooding pastures as well as to allow rain to run off.

The Brière Regional Park—which Alphonse de Chateaubriant described in his novel about the marsh, *La Brière*—is the second largest marsh in France, after the Camargue, and long ago included saltworks. Numerous small saltworks still exist in the area, on the islands of Ré and Oléron in particular: Ars-en-Ré and Loix-en-Ré are saltworkers' towns (*sauniers*, as they are called locally); Sauzelle (on Oléron) is another of these.

The Atlantic coast of France has been home to other saltworks sites: for example, those in southern Brittany, Trinité-sur-Mer, and the Gulf of Morbihan (Falguerec, Séné, Lasné), which have been turned into bird sanctuaries.

THE BEGINNING OF *BÉATRIX*. Balzac chose the site at Guérande for the opening scene of one of the novels of *The Human Comedy*. *Béatrix* opens with the following passage: "France, and more especially Brittany, still has some few towns that stand entirely outside the social movement which gives a character to the nineteenth century."[11] Béatrix-Brittany; especially-entirely; outside-character: this beginning establishes the equivalences and tensions in which characteristic features of the romanesque can take root. A bit further, one comes across this refrain: "In Brittany, where the character of the people allows no forgetfulness of anything that concerns the home country" (2). Then, quickly advancing to the conclusion of the novel's opening, to the threshold of the story, as it was termed not too long ago (in the day of Charles du Bos), this threshold is marked out, decked as if with a banner, with this description (a paraphrase, I might note, of the lyrical purple passage that ends *The Wild Ass's Skin*): "Now and again a vision of this town comes to knock at the gates of memory; it comes in crowned with towers, belted with walls, it displays its robe strewn with lovely flowers, shakes its mantle of sand hills, wafts the intoxicating perfumes of its pretty thorn-hedged lanes, decked with posies lightly flung together; it fills your mind, and invites you like some divine woman whom you have once seen in a foreign land, and who has made herself a home in your heart" (8).

"A foreign land," and Balzac to describe it for us, in its mineral and animal elements, borrowing his colors from the Painter (Le Nain, La Tour, Chardin?) and his prose from the Administrator (Vauban, La Pérouse, Turgot): "The whiteness of the linen clothes worn by the *paludiers*, the salt-workers who collect salt from the pans in the marshes, contrasts effectively with the blues and browns worn by the inland peasants, and the primitive jewelry piously preserved by the women. These two classes and the jacketed seamen, with their round varnished leather hats, are as distinct as the castes in India, and they still recognize the distinctions that separate the townsfolk, the clergy, and the nobility" (5). I will return to the pictorial mention of the "whiteness of the linen clothes [*toiles*]" applied right to the whiteness of the painter's canvas. [The noun *toile* means both "linen" and "canvas."—Trans.]

"A foreign land," with Balzac to make a sculpture of it, like those carved likenesses on gravestones: features softened by the marble's fine grain, still as a corpse, its flesh desiccated by the twin bite of salt and time. Balzac describes it anatomically: "The arms, bereft of nutrition, have dried up and merely vegetate" (2). He describes it again, in the register of medieval sculpture, of the fantastic bestiary of grotesques and monsters: "Woodwork, now decayed, has been largely used for carved window-frames; and the beams, prolonged beyond the pillars, project in grotesque heads, or at the angles, in the form of fantastic creatures, vivified by the great idea of Art, which at that time lent life to dead matter" (4).

Of course, one is reminded here of *Notre-Dame de Paris*. In the same register, that of intellectual history, that Hugo initiated in a brilliant manner ("This will kill that" [This is a reference to book 5, chapter 2, of *Notre-Dame de Paris*, entitled "Ceci tuera cela." The claim is that the advent of the printing press will "kill" architecture.—Trans.]), Balzac sketched out with the flick of a quill the project for an industrial archaeology—"to industry, monuments are stone-quarries or saltpeter mines, or storehouses for cotton" (3)—after he had defined the nineteenth century as the century when socialism and the social burst into being.

Purified by its remoteness from everything, suspended in the past, the novel's setting is a blueprint, a geometric figure, a diagram. The author defines its borders with three clips of the scissors at the "apex of a triangle." He traces out and projects a geography onto a map, a geography made aseptic by the passage of time, mineralized by a devastating epidemic the remains of which are these desiccated, mummified bodies crumbling into dust. In so doing, he turns the setting of his novel into a blank sheet of white paper, ready for writing to be inscribed on it. The blank slate and the white page are mandatory prerequisites. The novel can sing out its tune only after this inversion, which makes ornament out of the very absence of ornament: "This rich landscape, so homelike, so little visited, with all the charm of a clump of violets or lily-of-the-valley found in the midst of a forest, is set in an African desert shut in by the ocean—a desert without a tree, without a blade of grass, without a bird, where, on a sunny day, the marshmen, dressed all in white, and scattered at wide intervals over the dismal flats where the salt is collected, look just like Arabs wrapped in their burnouses" (7). "A foreign land," granted. The novelist takes the path of time in reverse; like Cuvier, he moves from whitened bones to living flesh, from the Guérande peninsula—beyond the world, folded in on itself, outside life and society, a fortiori beyond the social—to the portrait of a woman in society, to Béatrix.

One will have noticed how, before entering his tale, Balzac set it nowhere, to be precise, placing himself at a considerable distance from the here (alluding to the Sahara, even to Indian castes) and the now (evoking centuries gone by). The space of the romantic novel is not the space of the real, though it mirrors that space. And the first European settlers in North America met with native people, at ease in their living space and enjoying a dependency on nature, which, to them, was the respected dwelling place of gods, too.

ONONDAGA, SUCCESS, AND DECAY.   Onondaga Iroquois leaders in 1654 brought the French Jesuit Simon LeMoyne to their salt springs. Onondaga Lake, four and a half miles long and a mile wide, is located in upstate New York, near Syracuse, 130 miles west of Albany. The French returned there in 1656. They built a mission. They showed the Indians how to boil brine to extract salt, and they demonstrated various uses of the mineral, to cure meat, for instance. But they left in 1658, displaced by a few other European trappers and settlers. The Iroquois, powerful, resisted any general settlement in the area. Sporadic exploitation of the salt springs continued, however. Production by two escaped black slaves in 1774 is documented. They sold the salt to trappers, who used it for treating and preserving beaver pelts.

The same year, 1774, Congress appointed a committee, chaired by Thomas Jefferson, for regulating land settlement in the thirteen colonies and territories. This committee established the grid system of land subdivision before settlement, with townships made of thirty-six sections, each one mile square. Surveyors were charged to report natural resources, including streams and minerals, under nineteen headings.

The Land Act of 1796 included a provision for salt reservations, the intent being to avoid monopolistic takeovers. By that time, the Onondaga—one of the five nations making up the Iroquois Confederacy—had contracted a treaty (1778) with the state of New York. This treaty transferred to the state twenty thousand acres around Onondaga Lake for the purpose of salt manufacture, to benefit in principle the Indians and the white settlers jointly. Thus the salt springs came under the jurisdiction of the state of New York.

In 1788 Major Asa Danforth and Comfort Tyler, Revolutionary War veterans, settled in Onondaga on land given to them in lieu of salary by the newly formed government of the United States. They were joined by a few others—Nathaniel and Deacon Loomis, John Danforth, Thomas Gaston, Hezekiah Olcott—and the group was soon exploiting the salt springs. By 1797 the Onondaga area supplied almost all the salt used in the United States.

In 1809 the town of Salina organized itself. It took this name, already given to the area in 1797, when the state had taken control, in recognition of its main resource and livelihood. By 1824 the village of Salina had one hundred dwellings and sixty saltworks, with 1,814 inhabitants and 484 schoolchildren. Other saltworks operated in the nearby village of Liverpool. The next year, 1825, the Erie Canal was completed. This, together with the opening soon thereafter of the Oswego Canal (1828), made the two villages, Liverpool and Salina, extremely prosperous from their salt industry. A companion industry also mushroomed: coopering, that is, the manufacture of barrels as containers, for salt, in this case. Coopering was the near-exclusive province of German settlers.

In 1824 Salina already had an annual production of half a million bushels of salt for an average price of 12.5 cents a bushel (a bushel is 35.2 liters). The brine was quite rich in salt; a gallon (3.785 liters) of spring water would yield one pound (454 grams) of salt. Boiling was performed in kettles holding 100 to 150 gallons of brine, aligned in blocks counting up to seventy-eight kettles. It went on twenty-four hours a day. Hundreds of cords of wood were burned every day. When coal came to replace, indeed had to replace, depleted wood as fuel (by the 1820s the saltworks had burned out the entire local wood supply), about twelve tons of coal were needed daily to keep the fires raging. The equilibrium with nature that the Iroquois had achieved was already a fading memory. The cost of importing coal from Pennsylvania was starting to price Onondaga salt out of the market. By 1864 solar evaporation—the salt resource was still there and businesses do not let go easily—had taken over. It had become the primary means for salt production, even though it was constrained to the sunny season, from May till the end of October. At the height of this new mode of operation, fifty thousand evaporators were in action. Nevertheless, salt production went into slow decline from competition by cheaper sources of salt in the western states; it was finally brought to a close in 1926.

In 1884, however, the Solvay process for manufacturing soda (sodium carbonate) from salt and lime (calcium carbonate) had been introduced at Onondaga Lake. The by-product is calcium chloride. As is the rule in such coproductions, the demand for calcium chloride was less than that for soda. For many years, the unwanted by-product was dumped into the lake: a total of five hundred tons. Then an industrial accident occurred. In 1901, the company had begun storing waste slurries in diked marshes along the shoreline. On Thanksgiving Day 1943, the dike gave way. The brine flood drowned twenty houses. In 1953 Allied Chemical deeded four hundred acres of waste beds to the state of New York for one dollar for use as state fairgrounds; in exchange, the state agreed to drop claims against Allied for the 1943 waste spill.

Today, Onondaga Lake has become a chemical cesspool. A layer of calcium carbonate three to four feet thick lines its bottom. Whereas upstate New York is situated in a belt that experiences acid rain (mostly from coal-burning power plants), Onondaga Lake does not carry acidic waters; far from it: it has a pH between 7.6 and 8.2 (7.0 is neutral), that is, its waters are strongly basic, much like effluent water charged with detergent soap from a washing machine.

Mercury from the electrodes of the chloralkali process, which is also drawn to the area by the availability of cheap salt, is another, very toxic pollutant of Onondaga Lake. From 1946 to 1953 almost eleven pounds of mercury entered the lake daily.

The present estimated cost for a cleanup of Onondaga Lake is six hundred million dollars. Had the Onondaga Iroquois in the seventeenth century foreseen that the greed of the white man would bring about such an ecological disaster?

DESALINATION OF SEAWATER. The Guérande saltworkers separate salt from water and harvest the salt. But, conversely, how is one to obtain drinkable water, to get fresh water, from salty water? The apostle James exclaimed: "Can a fig-tree, my brothers, yield olives, or a grapevine figs? No more can salt water yield fresh" (James 3:12). Science, extended in the case in point by technology, is accustomed to such skepticism: after all, can't one define science as that which comes along and contradicts common sense? Certainly, nowadays many procedures that falsify James's apparently sensible claim have come into existence, and, indeed, if the apostle had lived in a climate less mild than that of Palestine, the same common sense might have led him to the notion of seawater desalination.

In the Arctic, as in the Antarctic, the ice field is made of fresh water, and yet it consists of water from the ocean that has frozen at the surface. Frost in refrigerators, produced from the water in the fruit and vegetables they hold, yields distilled water when melted. In like manner, rainwater, firn snow, and thus glacier ice are all fresh water. In other words, the evaporation of ocean water under the joint action of the sun's heat and the wind produces fresh water almost entirely devoid of dissolved salts. In this, nature provides us with a model. One of the desalination processes most used at present consists of heating seawater in a cooker and distilling it.[12] Once the steam has been condensed, one can collect water that is very pure. To get a clear idea of this, starting with seawater with a salinity of 35,000, measured in ppm (parts per million), the distilled fresh water contains no more than a minute amount—only between 1 and 50 ppm—of salt.

A variation on this is the process of multiple effect distillation. This makes use of the fact that in a partial vacuum water vaporization occurs at temperatures considerably lower than 100°C. Yet another variant on the distillation method, and also another thermic desalination process, is flash distillation. If one heats water to 100°C and keeps it under pressure, before introducing it into a closed container in which one has created a vacuum, one observes a very quick change:

the liquid is instantly transformed into steam. These two other thermal processes, multiple effect distillation and flash distillation, also produce water that is very pure (1–50 ppm of salts).

The other desalination processes, less effective with respect to salt content (it remains between 1 and 2 percent of its initial level, thus between 350 and 700 ppm), use semipermeable membranes. These membranes, made of synthetic polymers, let fresh water pass through but trap the salts it contains.

In the process of reverse osmosis, seawater is forced through the separating membrane at high pressure, a pressure higher than osmotic pressure, hence the expression "reverse osmosis." A fairly strong stream of water sweeps across the membrane so as to remove salts from it continuously, thus preventing them from building up on and clogging the membrane's surface and thereby permitting its nearly uninterrupted functioning.

In electrodialysis, an electrical field causes the ions to migrate, since they carry electrical charges, and these are trapped by means of membranes designed for this purpose. Most processes using membranes employ electric pumps to ensure water flow.

The output of the various processes is between 15 and 50 percent: 100 cubic meters of seawater must be treated to extract 15 to 50 cubic meters of potable water. No matter what method is chosen, the desalination of seawater remains a costly undertaking. The amount of energy required ranges from 2,500 to 12,000 kilowatt hours to produce 1,233.5 cubic meters of fresh water: this represents an expenditure of between $1,500 and $2,000. Existing production units provide between 500 and 1,000 cubic meters a day. This is the case, for instance, at the facility installed on Lavan Island in the Persian Gulf. And the plant in the city of Santa Barbara, which uses reverse osmosis to produce approximately 10 million cubic meters of fresh water annually, does so at a cost of about $2,000 per 1,233.5 cubic meters.

This shows, at the municipal level, the role of government in supplying an essential resource. In centuries past, the social contract was such that government, whether from tribal chieftains or heads of nation-states, would provide salt to the people. But at what cost? A horrendous one, as the next chapter will show. Too often, a callous, greedy ruler could not resist filling his coffers by levying a heavy tax on salt. What could the people do? They needed the salt. They would pay the tax, as long as it remained more or less tolerable.

four **abuse of power**

This fourth chapter examines the control by rulers of the distribution of salt to the people. It treats the taxation of salt and the rebellions it provoked, as well as the spread of this violence—whether it stemmed from the state or from the consumers—with the colonial dominion Europe granted itself from the sixteenth to the twentieth century.

The chapter opens with a general thesis of an anthropological slant: in many human groups, the technology for producing a consumer good is responsible for the founding of a culture. The argument proceeds with the reminder that the borders set between countries are imposed by the clash of two forces of brutality. But it is fitting to qualify this assertion by paying attention to a dialectical suggestion: a saying, uttered by Aristotle, that indeed concerns salt, referring to salt drawn from coastal marshes.

Venice, too, is born of the salt marshes, like Aphrodite from the wave. A brief history of the Republic of Venice teaches how a contingent fact, the refuge of its inhabitants in an insalubrious, marshy area, was turned by them into an asset. The asset became a domination when the marshland was developed into a salt production center. The domination in turn became an empire when Venice realized that it could emancipate itself from its local production operations by seizing competing producers nearby and overseas. The republic graduated from salt imperialism—which in the meantime had expanded to include numerous commodities issuing from the eastern Mediterranean and the countries of the Orient—to monopoly. This monopoly prevailed not only over large parts of the Mediterranean but also at the very heart of the republic: various influences of the Venetian salt monopoly on the very organization of the city-state are evident. These range from large-scale public works (beginning with the containment of the local rivers whose flooding was a serious threat to the saltworks) to a strong and coercive political regime and a city design based on districts divided into regular blocks, modeled on the topography of salt flats, with their grids of geometrical design.

But Venice, in spite of its hegemonic will, was obliged to battle other rival imperial powers—whether the Genovese (in production and trade) or the Dutch (in trade alone)—themselves also tempted to monopolize the delivery of salt.

Modernity thus roots one of its origins in salt. Salt production and trade provided the model for a colonial type of economic development, in which an essential good is monopolized by a country then in a position to fix its price, by force if need be. Moreover, the trade in salt has exerted a lasting influence on ways of thinking (in passing, note that it is one of the historical roots of anti-Semitism).

The Dutch Revolt, which took place in the Netherlands at the end of the sixteenth century, illustrates salt's relationship to the colonial type of subjugation and, conversely, to the struggle for independence from an occupying force. It entwined a religious war, a struggle for national independence, and a social revolt. The blockade by the Dutch of the Iberian salt production centers caused the bankruptcy of Philip II's Spain, in spite of its importation of gold—which disappeared with the speed of the fairy variety—from the Americas. The Netherlands' war of independence demonstrated the rise in the strength of its naval force, thereafter the leading navy in the world.

And what was happening in France during the same period under the ancien régime? The French people were suffering from the gabelle, a salt tax all the more unjust for penalizing the poor and that, further, discriminated between provinces in the kingdom. Made heavier by Richelieu, who relied on the revenue to finance the king's military expeditions, this tax became synonymous with royal power. Little by little over the course of centuries, the king arrogated the salt tax for himself. It rendered costly a product that was of the highest necessity, one that was abundant to boot. Ways of thinking retain a trace of this: one still speaks today of an "addition salée" [Literally, "a salted bill," a bill or check that is too high or padded.—Trans.]. In addition to everything else that was irrational about the salt tax—Vauban, a military architect and a visionary statesman, courageously denounced it—the brutal repression of the trade in contraband salt did not succeed in eradicating it.

The question then comes up: isn't every tax, because of its salty origin, stained with an intrinsic violence? Furthermore, the farming-out of taxes still endures. Tax farms still compare, even in their abuses, to those of the farmers-general [Fermiers généraux was the name for the wealthy and powerful private tax collectors in France before the Revolution.—Trans.] under the ancien régime. They persist today, for example, in the form of fees levied by corporations for water or highway use.

Other relics attest to the suffering endured by saltworkers over the course of history, whether in the mines or in the evaporation chambers of the saltworks. To visit today the Wieliczka mine or the utopian city of Chaux, designed by the ingenious architect Claude-Nicholas Ledoux but unfortunately never completed, is to witness a cold brutality. The domination in the collective psyche of salt's rarity and high cost is also reflected in sayings and proverbs from the most diverse cultures, from Ireland to Japan.

This chapter closes with an overview of a major episode in contemporary history: the Indian insurrection led by Mahatma Gandhi against the salt monopoly and the salt tax arbitrarily and abusively imposed by the English. Gandhi's 1930 salt march gave the signal for the struggle for independence; it was also innovative in opposing violence with nonviolence as an absolute weapon. Its lesson inspired numerous other movements for liberation from colonial domination, such as those carried out later in the century by Gamal Abdel Nasser and Martin Luther King Jr.

TECHNOLOGY AND SOCIAL STRUCTURE. According to anthropologist Mary Douglas, consumer goods are a form of communication, since they give rise to barter or exchange. Until the time of Gutenberg, they were the chief means of communication between human groups. Forms of communication in their turn dictate social structure: "The meanings conveyed along the goods channel are part and parcel of the meanings in the kinship and mythology channels, and all three are part of the general concern to control information."[1] And goods are the guarantors of social relations, which they help to establish.

Salt is exemplary in this regard. Several centuries before our era, sheep tracks used for transhumance already coincided with the paths used for salt distribution that went from the Mediterranean coast to inland mountain zones such as the Mercantour or the Queyras areas in the Alps or the Aubrac plateau in the South Central Massif of France.

Salt from the salt marshes was produced during the fair season, at the time when the herds of sheep were stationed in high mountain pastures. In autumn, when the flocks came back down to their winter grazing lands (at the mouth of the Rhône river, among others), mule convoys carrying salt left in the opposite direction, toward the heights, by way of river valleys such as the Durance, Var, Tarn, and Lot. In the spring transhumance, the flocks left the winter meadows for their high pastures, accompanied by supplies for the shepherds and their animals: salt-cured foods, salt blocks for licks, and salt for human consumption, whether consumed directly or indirectly (in cheeses). The same itineraries served for carrying wood—indispensable for boat building—to the rivers and the coast.

Did the barter of salt for wood—to limit ourselves to this economic exchange—influence, or even determine, a certain type of family and social relations? Did the interrelations of corresponding occupations—saltworkers and woodcutters, sailors and mule drivers—leave behind traces in language or in social practices?

It is reasonable to infer an economic and cultural interdependence between coastal populations, settled in salt marsh locales, and mountain peoples, inhabiting the forests of deciduous trees exploited for their wood. One may likewise conjecture an extended, even clan-based, family structure: communication through exchange of goods is facilitated by existing kinship bonds: when the mule driver and the woodsman are cousins of the fisherman and the saltworker. Moreover, the seasonal nature of most of these activities requires men to be absent for rather long periods. It is thus conducive to strict rules regarding their wives' fidelity. To this day, Corsica still preserves such a social organization: the vendetta rages between clans whose enmity is rooted in an initial act of aggression, and family solidarity protects men sought by the authorities, provides them with hideouts in the mountains. Salt smugglers in modern times, during the seventeenth and eighteenth centuries, took full advantage of the marginal nature of these very old affinities between the coast and the forested massifs.

The European expansion, beginning in the fifteenth and sixteenth centuries, was buttressed by salt technologies. The salt curing of fish, herring, and cod was one such innovation, with significant social consequences. Shortly after the discovery of America, European fishermen, the Portuguese in particular, went to fish cod in the seas off Newfoundland, letting the fish dry on the shore before bringing it back, preserved in salt, to Europe.[2] This salt cod was an abundant source of protein for coastal populations from the Pillars of Hercules to Brittany and the North Cape.

Deep-sea fishing itself was based on the preservation of proteins by means of salt: on the salt curing of hams, the aforementioned salting of cod, and the salting of herrings.[3] The discovery voyages of the great navigators made use of these food preservation techniques, which amounted to a sort of canning industry *avant la lettre*: the European expansion of the eighteenth and nineteenth centuries, the settlement colonies, the slave trade in black Africans, and the planting of sugarcane in the Antilles were all consequences of salt technologies and thus of the same model of economic prosperity embodied in the mercantile cities of the Mediterranean and the Hansa.

Can one refine this model and advance the hypothesis that sharing a single technique for producing a consumer good creates a common culture, which would permit a better understanding of some social facts? Certainly, in the sixteenth century, for example, if one considers as a whole those European societies that obtained their salt from salt marshes, from the Adriatic to Brittany, one easily observes that the ocean-going navigators were the Venetians, the Genovese, the Portuguese, and the Bretons. However, this explanation is too hasty: if Genoa and Venice were powerful mercantile cities, why was it not the same for Nice or Brest? How is it that the islands endowed with salt marshes and saltworks, such as Sardinia and Cyprus, did not experience economic development and urbanization at the end of the Middle Ages comparable to that of Genoa or Venice, even though the material factors seem a priori alike?[4] How was an inland city like Florence able to rise in power during the same period as Genoa and Venice and in a comparable way?[5]

If Genoa and Venice devised political systems for themselves that were analogous, down to the doges, why was Florence an exception? Can one legitimately compare the present-day economic success of Singapore with that of Venice at the end of the Middle Ages? How is it that so many mercantile maritime cities—from Venice to Singapore, as a matter of fact—have opted for authoritarian regimes when trade seems rather to cause democracy to flourish, as in ancient Athens?

The onus is on historians to answer these questions. Other fascinating questions can be asked of anthropologists. These concern representations: how is it that in the collective imaginary as it unfolds in mythology, the story, the novel, the melodrama, and the opera, the character of the saltworker appears so rarely? Though each language has proverbs and sayings on salt, samples of which are sprinkled throughout this work, the method for its production has remained obscured, as if repressed, in collective representations. Could this be the legacy of a bygone social structure, lost in the dark of time, when saltworkers would have constituted a distinct social class of inferiors, segregated from the rest of society by marriage prohibitions?

But let us move from prehistory to history, so as to consider the birth of nation-states and their relation to salt and its production and trade. Autocracy becomes so tempting, since it feeds on some of the basic needs of one's fellow human beings.

NATIONAL SOVEREIGNTY.   Some former international borders have vanished from collective memory. Only a historian would remember nowadays, while driving along the Riviera, all the tolls the local lords had the mule drivers carrying salt to Cuneo and Turin pay. Who recalls the 2 percent tax imposed by the Republic of Genoa on all sea transport of salt or other goods crossing its sector of the Mediterranean, then named the Sea of Genoa? Since the 1860 annexation by France of Savoy and the county of Nice, we have lost the memory of such earlier borders. Who recalls, likewise, the location of the border between France and Franche-Comté prior to its incorporation into France in 1678 (the date of the Treaty of Nijmegen)?

Such a glance in retrospect neatly points up the artificiality of today's borders between European states, of the need to establish the Common Market, and thus of the inevitability of the project of building the European Community.

Since this very question incites violent and impassioned reactions, one ought to reflect on the erasure of historical facts that were vivid and pressing in their time. One is tempted to label as archaic notions of national borders and sovereignty; granted, they are so, to some degree. But consider present-day Africa as a counterexample: the scrupulous respect of borders "imposed by colonization," to use the accepted term, is a rampart against Balkanization and the conflagration of the whole continent into tribal wars.

Plainly, borders result from land ownership. As soon as humankind discovered agriculture, land ownership became a fundamental juridical concept. As a consequence, boundaries acquired major importance to human groups set either to seize the riches of other groups or to resist such attacks and repel invaders. The history of the last two millennia is to a great extent the chronicle of the establishment of territorial communities, of the borders that demarcate them, and of the transgressions, most often bellicose, of these borders.

Strabo, a Greek geographer who lived in the first century before Christ, thus reported: "The Autariatae were once the largest and best tribe of the Illyrians. In earlier times they were continually at war with the Ardiaei over the salt-works on the common frontiers. . . . They would agree to use the salt-works alternately, but would break the agreements and go to war."[6] But, to an equally great extent, a border is the internalized awareness of belonging to a community. Cultural identity was forged during antiquity—Greek and then Roman—with definition of such partitions: hither, people recognized as one's own kind, sometimes as one's equals; yon, barbarians with bizarre and vaguely reprehensible habits. Thus the border does not simply mark out ownership (everything west of a certain line belongs to the Blues, everything east of that line belongs to the Reds); it is also the protective Maginot line and the threshold of xenophobic rejection, the buffer zone. Can cultural identity be built in any way but the polemical?

An apparently rational answer—arguing that cultural identity is rooted in a common language, in a joint religion, in a single set of shared customs and juridical rules—does not wash. The diversity of dialects, family structures, weights and measures, and even the wars of religion did not prevent France from existing centuries before the Convention came about and imposed its unification on it. And, though there is no room here to elaborate on this further, rather clear connections link the seemingly positive notion of cultural identity with the plainly negative notion of racism. Cultural identity is one of the garments of naked power, of the brutality of tyrants, often great generals, whether Julius Caesar, Louis XIV, or Napoleon.

Sovereignty, in long bygone times and for most of the Middle Ages, was associated with the tripartite division of society into *bellatores, oratores,* and *laboratores*: warriors, priests, and peasants. Sovereignty flowed from property ownership after an implicit social contract divided society into these three classes. Peasants supported through their labor—in kind or through various payments—warriors (and priests), who in exchange guaranteed their physical security and that of the community's assets. The medieval fortress, ringed by fields, protected by high ramparts behind which all found shelter when a band of marauders swept through or when under attack by an organized enemy troop, is emblematic in this regard.

Seigneurial power (further expanded but still unchanged in its latest embodiment, the twentieth-century welfare state) also included the moral obligation of paternal benevolence toward subjects: besides the duty to attend to the physical security of people and goods, this meant ensuring their subsistence, guaranteeing them the supply of indispensable commodities such as salt.

To sum up, the nation-state cannot be conceived without wars in which it confronts its neighbors; every border presupposes hostile forces on both sides. But this assertion needs refining, and reference to another saying comes in handy here.

THE PROVERB OF THE MARSH PURCHASE.   "Buying the marsh with the salt" is a proverb quoted by Aristotle.[7] The context is that of the orator having to choose between two opposed presentations of a single concept. The examples Aristotle gives are those of the well-bred man, proud of his education but also envied for having thrived because of his upbringing, and the public figure who, if he says what he thinks just, invites the hatred of his fellow citizens, whereas if he makes unjust comments, the gods will hate him.

This gloss helps us understand the proverb Aristotle quotes: it likewise has two opposing sides. Salt, with its positive meanings (the practical use, the symbolic relation to conviviality, the humor of Attic wit [*Le sel attique* is the French for "Attic wit."—Trans.]), symbolizes the good. The marsh is its counterpoise, with its negative connotations (treacherous ground impossible to farm and unsuitable for grazing, with malignant malarial exhalations); the marsh symbolizes evil.

The philosopher then draws a number of lessons from this proverb of the marsh purchase: since good and evil are two sides of a coin, they can be hard to distinguish. Likewise, one may not make the same assertions in the two spheres, the private and the public. A certain mental habit is called for: on being confronted with only one element in such a double-sided discourse—for example, when listening to a paradox—one should have the shrewdness to infer, by turning it around, its missing complement.

This incisive rhetorical tool applies to the argument I have just presented. Nothing in fact prevents two neighboring nations from coexisting peacefully, even if, over the course of history, this has occurred relatively infrequently.

VENICE.   Some tourists of today's Venice take the boat to Torcello island. The island was settled in the fifth and sixth centuries by the Altinians when they were driven off the mainland by the Lombard advance. They made it an important political, religious, and economic center. It was the kernel from which Venice was born.[8]

Today's visitors to Torcello, come to see the two back-to-back medieval churches and the Byzantine mosaics of the larger one, are entranced by the tranquility of the setting. They have the feeling of being far removed from the world, in a place of spirituality, molded by history. Sculpted stones repose on the ground.

Around the year 1000, the sector from Torcello to Murano—the island now occupied by glassmakers—was one of the first Venetian lagoons where saltworks were installed. As I summarized above, it can even be maintained, following Jean-Claude Hocquet and Michel Mollat, that the Republic of Venice built its economic prosperity and commercial dominance on salt and its trade.[9]

The presence of the saltworks in the lagoon is confirmed from the tenth century on: at Lido Sant'Erasmo in 958, in Chioggia Minore in 991. Then, later, saltworks were installed in Equio (1022), Murano (1015), Lido Bovense (1038), and in Venice itself (1046).[10]

As is well known, Venice became a great maritime power. The city prided itself on its mythical relation to the marine element, symbolized in the annual ritual commemorating the wedding of Venice to the sea. In fact, salt flats mediate between sea and sun; they use solar energy to evaporate seawater. Driven back to the shore of the Adriatic, the Altinians, the lagoon's first inhabitants, developed its raw resources, especially salt harvested in the saltworks. The future Venetians then exported salt to the backcountry and initially bartered it for the grain they needed. Adversity had thrown early Venice into the marshes, where the Venetians made salt and made it the seed for their future prosperity.

Since the mainland cities henceforth depended on Venice for their supply of salt, this city gave itself a position of monopoly first on salt production and then on the sale of this precious commodity. To this end, Venice waged numerous wars: it aimed to eliminate the competition from other northwest Adriatic cities—such as Comacchio, its great rival in salt production—so as to seize or destroy their saltworks. In addition, Venice entered into profitable commercial treaties, often in the aftermath of victorious wars, making the cities of Paduan Italy depend exclusively on Venice for their salt supply.

The height of salt production in the lagoon took place at the end of the twelfth and the beginning of the thirteenth centuries. Jean-Claude Hocquet has made a systematic study of the manuscripts: they record saltworks covering as many as 119 *fondamenti* (units of area measurement). But in 1348 only 37 *fondamenti* were left in the south lagoon, in Chioggia (where salt production would continue until the twentieth century). Moreover, these saltworks extended over large areas: for example, the saltworks at Cona da Corio measured about two hundred thirty paces long and one hundred or so paces wide, in other words, an area on the order of seven hectares.[11]

The Venetian saltworks were structurally fragile, at the mercy of the sea's erosion of the Adriatic's sandy shores, the hazards of storm or flood, of epidemic striking the saltworkers, and above all the overflows and course changes of the three rivers with their estuaries near the lagoon, the Brenta, the Piave, and the Sile.

Large-scale public works were indispensable for diverting the courses of these three rivers. Therefore Venice very early on devised for itself a strong political system. Indeed, an authoritarian rule, enjoying a lasting period of power, allows for the carrying out of large-scale public projects and for their subsequent upkeep. In the thirteenth century, Venice once again innovated by establishing the gabelle: this salt tax results likewise from a strong political power that in this way gives itself increased economic power. The establishment of the salt tax marks the transition, to borrow Michel Mollat's periodization, from an initial phase of salt production in the saltworks, with monasteries as owners of the production sites (from the ninth until the thirteenth centuries), to an intermediary phase of state production, when the Venetian state seized this highly lucrative property.

The third phase is that of trade in salt, shipped throughout Europe from the Mediterranean and Atlantic saltworks. In the fourteenth and fifteenth centuries, Venice, which hardly produced salt locally anymore, continued to control its production in numerous and at times remote foreign saltworks; it had salt brought to it for reexport from points of origin such as Tripoli in Libya; Zarzia in Tunisia; Cyprus; La Mata in Spain; Ibiza in the Baleares Islands; Chiarenza in Greece; Barletta in southern Italy; Pago, Spalato (present-day Split), and Zara in Dalmatia; and Piran in Istria.

Thus salt production and subsequently its trade provided the Venetians, from the first centuries of the republic's existence, with an economic blueprint: a monopoly position allows for rapid accumulation of wealth, since one retains control of the selling price. Venetian domination was imposed on the northern Italian cities to which Venice supplied salt. The republic thus learned the importance of coupling economic dominance with military superiority so that weapons could, if necessary, guarantee the integrity of the monopoly. Venice subsequently consolidated its mastery, if not its exclusive control, over the importation of spices, perfumes, fabrics, and other luxury goods from the Orient to the Mediterranean world. Its fleet gave it strategic superiority in the eastern Mediterranean, against the Turks, in particular. It would seem that the first chapter of its life, a salt-making one, served it as a lasting lesson thereafter.

One cannot resist visually comparing the plans for the saltworks (such as the texts have bequeathed them to us) with maps of the city of Venice: one finds the same grid pattern, the same branching structure of the primary, secondary, and tertiary canals that mark out the dry land areas, that is, the city's quarters. The model of the saltworks plainly determined the city plan of Venice, which in turn has had a lasting influence on other city plans, on that of American cities among others. New York City, for instance, beyond its past as New Amsterdam, was imprinted with an urban grid patterned after the Venetian saltworks. It was a New Venice, too.

In the seaports of all maritime countries, it used to be the practice, and possibly still is, that any merchant arriving there with merchandise, having discharged his cargo, takes it to a warehouse, which in many places is called the *dogana* [customs house], and is maintained by the commune or by the ruler of the state. After presenting a written description of the cargo and its value to the officers in charge, he is given a storeroom where his merchandise is placed under lock and key. The officers then record all the details in their register under the merchant's name, and whenever the merchant removes his goods from bond, either wholly or in part, they make him pay the appropriate dues. It is by consulting this register that brokers, more often than not, obtain their information about the amount and value of the goods stored at the *dogana*, together with the names of the merchants to whom they belong. And when a suitable opportunity presents itself, they approach the merchants and arrange to barter, exchange, sell, or otherwise dispose of their merchandise.[12]

This description of commerce, of brokerage, of customs duties, in short, of some of the tools of capitalism, has hardly aged. It is nonetheless more than six centuries old, drawn from Boccaccio's *Decameron*, published in 1353.

Following the withdrawal of populations to higher ground and the disrepair of the roads—or the deliberate damaging of them to make them unpassable, after the barbarian and Saracen invasions—commercial Europe expanded—but limited itself as well—along the coasts alone; such coastal expansion endured for several centuries (until the end of the Mongol invasions, toward the middle of the thirteenth century). Several cities along this coastal chain made good use of their location in their bid for power between the eleventh and fifteenth centuries:[13] Venice, whose zone of influence was primarily the Adriatic Sea, its east and west coasts, and the eastern Mediterranean (including the Greek archipelago);[14] Genoa, its great rival, which dominated the eastern Mediterranean;[15] the Netherlands, which exercised its domination over the Atlantic traffic;[16] and, finally, the Hanseatic League, which beginning in the twelfth century united mercantile cities such as Hamburg, Lübeck, and Kiel, operating in the North Sea and the Baltic.

The transport of salt, and the more or less successful attempt to establish a monopoly on it, accounts for the power of each of these centers. Venice, whose own salt production was in fact no longer dominant, secured for itself control over production in other salines: in Chioggia, more to the south; in Cyprus, with its Great Salt Lake; and in Ibiza, in the Baleares. In the middle of the fifteenth century, its sales approached thirty thousand tons annually.[17] Genoa controlled production at the Hyères saltworks, which in the same period accounted for two thousand tons annually. The Dutch dominated the trade in Portuguese salt from Setúbal, as well as that of Bourgneuf salt and the output of other French saltworks on the Atlantic coast.[18] The Hanse marketed Lüneburg salt especially, which it transported as readily westward (to the Shetland Isles, for the salt curing of herring, to London, and to Bruges) as northward (Bergen) and eastward (as far as Riga, Revel—today's Tallinn—and Novgorod); it also supplied itself with salt from the Bay of Bourgneuf.

A particular colonial type of economy is linked to this transport in salt, that is, an economy of trading posts. It ensures that the boats (whether feluccas, brigantines, or luggers) don't make the return trip empty, so that there are also commodities for exchange, other products to replace the salt in the holds. These products include fish (Europe at that time derived the bulk of its protein from fish) and another important product associated with salt in this precapitalist economy: lumber, used especially to build the vessels. It is a colonial economy: vast territories are held, with no real need to see to their administration, simply by means of state control of the influx and outflow of essential market goods. On the import side, there was the salt, indispensable for nourishment, and on the export side, there were all the local products to choose from.

Because weather is unpredictable and can wipe out with rain the annual output from solar saltworks on the seashore, salt underwent enormous price fluctuations. At times an inexpensive raw material, though one burdened with heavy transport costs, it could also turn into a product of sudden rarity. Therefore, whether salt was the basic commodity in a given maritime trade or merely return freight often remains an open question. This latter scenario, it would seem, applies to Hanseatic commerce in the fourteenth and fifteenth centuries.[19]

Thus the trade in salt is the canvas on which the economic history of Europe over the course of three or four centuries is traced. If Venice was the pioneer in combining military force, political power, and monopoly control of a commodity, this economic-political model radiated far and wide, beginning in the twelfth century. But it is especially in the fourteenth century that one witnesses the establishment of a double domination—of lords and adventurous ocean-going merchants—responding to the increase in northern European salt consumption, occasioned by the preservation of herring and then cod.

Among the ranks of the first group stood eminent the count of Burgundy, the principal member of an oligarchy that controlled the great Salins production center in Franche-Comté.[20] A transformation had taken place in the second half of the thirteenth century. Prior to that period, the economic landscape had been one of production dispersed among numerous competing centers. This pattern became simplified through the creation of a very small number of oligopolies in which the counts of Burgundy took the lion's share for themselves. During the same period, the lords, who included the dukes of Burgundy and more specifically the counts of Flanders, levied a new tax to enrich their coffers through a revolutionary technological innovation based on salt and its long-distance transport: the salt curing of herring.

As for the adventurous fourteenth-century merchants who would, for example, pick up salt in the Bourgneuf Bay and bring it back to Hanseatic ports and the trade centers of the North Sea and the Baltic, they established with this a model followed thereafter for colonial expansion into distant lands in the sixteenth-century voyages of discovery.

Military power is indispensable for dominion over a zone of influence. It is constantly put into question by skirmishes at its edges: Venice had to fight on the east against Byzantium, against the Turks, against the Arabs, and on the west against the Genovese.[21] Economic war takes many forms, from pirating to naval battles, from blockades to the long-term supply contract. In the fifteenth and sixteenth centuries, there occurred the seizure of 1449, when the English captured 108 ships from the Hanseatic fleet;[22] there was the contract with Genoa, allotting more than 4,000 tons of salt per year to Francisco Sforza, duke of Milan from 1495 to 1535; there was also the anti-Venetian coalition set up in Cambrai in 1508,[23] as well as the hostility of the Lombards, requiring Venice to stockpile its salt, so that between

1523 and 1530 Venice accumulated an enormous quantity—50,000 tons—in its warehouses.

Commerce in salt not only underwrote anti-Venetian sentiment, it also served as one of the roots of anti-Semitism.[24] Owing to the presence of salt outcrops on the shores of the Dead Sea and the importance of salt in Jewish culture and religion, Jews played a major role in the production and sale of salt. Already in the tenth century, Jews operated the salt mines near the town of Halle (the name means "salt") in Germany. In the twelfth and thirteenth centuries, Jewish entrepreneurs secured salt concessions in Spain from the sovereign. In the fourteenth and fifteenth centuries, Jews gradually entered the Polish salt trade, to such an extent that in the sixteenth century a portion of the Polish petty nobility embarked on a campaign to try to wrest away this privilege. In the seventeenth century, Jews also ran the Dutch and German salt businesses (from Setúbal to Gdansk).

To sum up: during the precapitalist period from the twelfth to the fifteenth centuries, the salt trade served as the seed for modern times. In European history, the waning of Spain as a major power and the birth of the Netherlands in strife were proof of the struggle for supremacy over the salt trade.

THE DUTCH REVOLT.   The Dutch Revolt, the first act in the armed struggle of the people of the Netherlands for independence from their Spanish occupiers, took place in the context of the international trade in salt. At that time, salt production centers were very geographically restricted, exactly as today's petroleum production is to a large extent tied to the Persian Gulf.

In the second half of the sixteenth century, the salt supply for the Low Countries, as well as for the Baltic countries,[25] came from the Iberian peninsula (especially from Setúbal in Portugal) and from the Atlantic coast of France (Brouage and the Bay of Bourgneuf). This salt was needed for the curing of meat and fish (herring). The Dutch, primary users of sea salt, also dominated its sea trade.

In the expression "Guerre des Gueux" [Guerre des Gueux is the French name for the Dutch Revolt. *Gueux* is a noun meaning "beggar, tramp, wretch; knave, rascal."—Trans.], the last word is a bit misleading. In fact, the revolt was started by the Dutch petty nobility, which depended for support on a popular revolt. This social class of country squires had converted en masse to Calvinism after 1650. They held armed assemblies, which made them aware of their military might. In October 1565 Philip II of Spain touched off a crisis by refusing any compromise arrangement: the Inquisition was to continue its work in the Netherlands, and the edicts against heresy would be enforced. On April 3, 1566, two hundred cavalrymen (who called themselves "*gueux*," i.e., beggars) from this Protestant petty nobility, carrying a petition bearing four hundred signatures, appeared in Brussels before Margaret of Parma, who ruled the Low Countries.[26]

Their action coincided with the discontent of the lower classes: the Nordic Seven Years' War between Denmark and Sweden (1563–1570) had shut off the Baltic Sund to traffic, and the textile and naval industries suffered in consequence, which provoked a serious economic crisis. Moreover, the winter of 1565–1566 brought a famine that the popular masses blamed on the government and speculators. In August 1566 commodity prices soared, and the revolt broke out on August 10 (269–70).

Philip II's reaction, in spite of his characteristic indecision, was to attempt to nip the insurrection in the bud. To this end, he deputed the duke of Alva to head a military peace-making expedition. But this took some time, and the duke of Alva did not arrive in Brussels until August 22, 1567 (270–72). His policy of repression, coupled especially with the imposition of a general 1 percent land tax, provoked an outcry. Repudiated in part by Philip, Alva attempted in vain to negotiate with the states-general of the Netherlands, while William of Orange took over the leadership of the diplomatic and military resistance to the Spanish. As for the duke's attempted military reconquest, it failed for lack of sufficient naval support (274–75).

The feature of the Dutch Revolt of relevance here is the military control of the Dutch coast, which then included what is today Belgium. The Dutch troops consisted of the Gueux des bois (soldiers of the woods), on land, and the Gueux de mer (soldiers of the sea), serving on the Dutch fleet. This last group took control of the sea and the coast. In April 1572 they seized the little port of La Brielle (Brill), on the island of Voorne, and then undertook the methodical conquest of the inland cities: Holland and Zeeland fell into their hands (273). This territorial seizure was coupled with naval hegemony: starting from the ports of Emden, La Rochelle (in French Huguenot hands), and England, the Gueux de mer blocked all trade between Spain and the

Netherlands. In this way, they prevented entry into the Netherlands of salt coming from Spanish saltworks. This caused Andalusian salt exports between 1576 and 1578 to plummet. The disruption of the trade carrying Atlantic salt through the Sund Strait to the Hanseatic cities was felt as far back as the Atlantic saltworks.

In the winter of 1572–1573, the duke of Alva attempted a massive and ruthless counteroffensive. Once again, it ran up against its own naval inferiority, and mass conversions to Calvinism caused an active minority of fanatics (so termed by the Spanish Catholics) to rise up against him (274).

During this period, for a number of reasons, first and foremost the naval blockade of salt exports from the peninsula, Spain headed rapidly for bankruptcy, which was declared on September 1, 1575. The soldiers of the Spanish occupation army in the Netherlands, whose wages were no longer paid, mutinied (275–76). This was the beginning of the end for the Spanish Netherlands. The Dutch Revolt ended in 1609 with the victory of the Dutch Calvinists.

The Treaties of Westphalia (1648) contained clauses to prevent and forbid from then on such salt blockades, which harmed every country involved. The episode contributed to England's resolute decision to give itself a strong and autonomous navy and thus to become a great maritime power. It also gave a lift to mineral mining on the Continent.

The Dutch Revolt combined elements of a religious war, a popular revolt, an embargo, and a blockade (in Lisbon and Setúbal in 1585, 1595, and 1598), as well as an economic war. Dutch naval power was a key element. It intervened to ensure the indispensable resupply of salt to the Netherlands. Since deliveries from the Portuguese saltworks were interrupted, it was absolutely necessary to find salt elsewhere. The Dutch had it brought from Bourgneuf, but in insufficient amounts.

It therefore became profitable to bring in salt from more distant places, first from the Cape Verde Islands and then, starting in 1595, from the West Indies.[27] In 1598 the Dutch salt merchants began to develop the vast deposits accumulated around the Araya lagoon, near Cumaná in Venezuela. Freighters specialized in such trade carried on the order of three hundred tons. From 1600 to 1606 one hundred such boats arrived to take on salt headed for the Netherlands. This new transatlantic traffic, which made the Netherlands a naval power unconfined to Europe (like the naval powers of Spain and Portugal before it), ended in 1609 when a truce was made and importation from Setúbal could resume. Over salt, countries went to war with one another. They also hounded their own citizens over its possession and trade.

THE GABELLE.   On July 25, 1785, J. Fournié of Saint-Martial, in the Albi area, was fined 200 livres (pounds) for contraband traffic in salt. Since he was unable to pay, to start with he had to undergo flogging as punishment: "We order that said Fournié shall be delivered over to the executioner of the high court, who, having stripped him naked to the waist, will lead him through the high street to the Lower Market-hall of the Convent, follow the market-hall to take the rue de la Fontaine, thrash him in the customary spot, and lead him to the Place du Pont where he will brand him on the right shoulder with a red-hot iron bearing the letter G." This sentence from the gabelle office in Ville-franche-de-Rouergue, recorded July 29, 1785, sought to be exemplary and terrifying punishment. Revolution has been provoked by less.

In fact, despite its atrocious nature, this sentence was also supremely ordinary. A fine of 200 livres was the standard rate applied to those convicted of contraband traffic in salt throughout the seventeenth century. That sum amounted to the average annual income of a common laborer, and it was thus impossible for the average person to have saved such an amount. His punishment thus amounted to being condemned to the galleys for life. The letter G with which Fournié was branded with a red-hot iron was the first letter of the word "*galères*" [*La peine des galères* was the punishment of being condemned to row on state galleys.—Trans.]: it symbolized his condemnation to penal servitude from then on.

The gabelle was always and everywhere hated. The salt tax was the prototype, in Western European countries, for direct and indirect taxes as we know them nowadays. The gabelle gave rise to an abrupt, arbitrary, and intolerable increase in the price of salt. This everyday mineral acquired the reputation of being costly, and memory of this has endured.

The gabelle, or at least its role as a tax on salt (to which the term *gabelle* was subsequently applied exclusively), serves as a sort of synopsis of the history of France: feudal powers clash; then one witnesses the gradually expanding ascendancy of the king of France; finally, there follows the necessary standardization—in opposition to numerous centrifugal forces of dispersion—as was enforced by the Convention.[28]

The first phase can be illustrated with the example of the license granted to the Lérins abbey, located on Saint-Honorat island off the coast of Cannes. On February 9, 1453, King René of Provence granted the monastery and its Cistercian monks the right to withdraw each year a certain amount (twenty-five *setiers*) [A *setier* is an obsolete measure for liquids and solids, equivalent to 400 pounds or one-fifth of a metric ton.—Trans.] of salt from the storehouses of Grasse, Fréjus, or others, depending on the monks' preferences. He confirmed this right in letters patent of June 14, 1453.[29] Prior to this, actually, the counts of Provence had granted the monastery—in return for an increase in religious services and for the protection provided by the tower—the right to take one *setier* of salt for each boat making deliveries to the storehouses at Cannes or Grasse. But the clever Provençals had foiled the monks by arguing that the tax to benefit Lérins was one *setier* no matter what the size of the ship. The story, of course, did not end there. One still finds formal complaints filed at the end of the sixteenth century by the Lérins monks claiming the twenty-five *setiers* of salt that, contrary to the agreements, had not been paid out to them by the gabelle tax farmers.[30]

The tutelage of the king of France gradually grew heavier beginning in the seventeenth century, and not only because of his aim for hegemonic centralism. The huge ransom for Jean le Bon had to be paid to the English, which vastly inflated taxes, following the Treaty of Brétigny and the 1360 order from the aforementioned King Jean. His decree was a continuation of the orders of Philip VI (1331 and 1343) that limited the sale of salt to royal storehouses alone and added to the price of salt the burden of royal rights.

This second phase of the gabelle is distinguished by great regional disparity in the tax levied. This fiscal inequality—in a tax on a commodity of the utmost necessity and thus proportionately more oppressive and unbearable for the common people, those of the third estate—worsened. Toward the middle of the sixteenth century, in 1548, there was in fact a bloody antigabelle insurrection in Angoumois and in Guyenne. The king of France had to make concessions. He accepted a plan whereby these provinces and their neighboring territories would permanently redeem their salt tax debt. These regions

(Poitou, Aunis, Saintonge, Guyenne, Angoumois, Limousin, Marche, and a part of Auvergne) were thereafter exempt from the tax and thus referred to as "redeemed provinces." In the other provinces, great inequalities in statutes persisted, in spite of an effort by François I at uniformity. The lands exempt from the gabelle, where trade in salt was free, were located on the borders of the kingdom: Artois, French Flanders, Hainaut, Béarn, Navarre, and Brittany. The *pays de salines* derived their salt from the saltworks of Franche-Comté and Lorraine and enjoyed no gabelle.[31] The monopoly on sales to profit the king lasted until the Revolution in the nonredeemed provinces. Elsewhere, after Colbert's great ordinance of May 1680, the statute was again modified. The greater part of the kingdom was subject to the heaviest tax, the *grande gabelle*, or great gabelle. Besides the tax itself, this included a required consumption minimum: at least one *minot* (around a hundred pounds) per group of fourteen people over the age of eight. Salt used for salt-curing procedures wasn't included in that calculation and had to be purchased over and above that amount.

The less burdensome *petite gabelle*, or lesser gabelle, affected all the southern regions (Lyonnais, Beaujolais, Mâconnais, Bresse, Languedoc, Roussillon, Provence, Velay, Forez, and the *élections* of Rodez and Millau in the *généralité* of Montauban). [*Election* and *généralité* are legal terms from the period of the ancien régime. *Election* means a financial district administered by an elected body. *Généralité* means a financial district directed by a *général des finances.*—Trans.] Salt was less costly there by a factor of almost two.

As a result, at the end of the seventeenth century a heavy and despised tax was in place, with great disparity in rates, regional variation, a massive and organized contraband trade, and corrupt tax collectors. The great variety in salt prices fed the contraband trade, in which men, *portacols*, as they were called—the man named Fournié, mentioned above, was one such *portacol*—carried loads of twenty to forty kilograms of salt.[32] Consumers were all the more given to fraud since the tax farmers of the salt depots, and their assistants, grew rich at their expense: it was common practice to fill a bushel without tamping down the salt so as to undersupply by about 12 percent.[33] At times, the tax collectors (customs officers) were corrupt and themselves engaged in contraband trade.[34]

The small number of salt production centers, coupled with the fact that everyone was a salt consumer, made the taxation of salt easy and the gabelle a ubiquitous tax of long standing.[35] Richelieu did not resist this all too easy means of raising new revenue: beginning in 1624, he gradually doubled the price of salt over the course of a little more than a decade.[36] The gabelle did not affect salt alone; tobacco and other goods were also subject to it.[37]

It is not surprising that this tax, so unjust, should have made a deep impact on the minds of the people and remained in the collective memory. By way of example, here are some figures: in seventeenth- and eighteenth-century Provence, production costs for salt remained stable, less than four *sous* [A *sou* is an old copper or nickel coin worth one-twentieth of a livre.—Trans.] per *émine* (about forty-five kilograms). The gabelle imposed on it made the sale price to the public rise to 14 times the cost of production in 1630 and then to 140 times in 1710.

In that same year, 1710, there occurred one of the countless examples of peasant rebellion that illustrate the interdependence of an iniquitous tax, contraband, and repression: in September customs officers confiscated the mules of the inhabitants of Gréoux and Saint-Julien who were engaged in contraband trading with the collusion of certain tax collectors. The customs officers, having taken refuge with their take at the Auberge du Chapeau Rouge at Vinon (in the Verdon area), were besieged on September 10 by a riotous crowd claiming back their livestock, since mules were one of their most essential work tools. On September 11 the customs officers were set free by the king's constables. From October 1710 to March 1711 the authorities investigated and proceeded to make arrests. The trial was held at Aix on April 24, 1711. In addition to eighteen convictions by default and seizures of goods, the court issued twelve other sentences: eight sentences to fines of one hundred livres, two to flogging with banishment and fines of three hundred livres, one to death by torture on the wheel, and one to the galleys for life. The aim of making an example of those convicted is evident.[38]

France definitively abandoned the gabelle only on January 1, 1946.[39]

In Swedish, the expression "en saltad räkning" designates what we call in France "une addition salée." Most languages have expressions of this sort, which derive both from the once-elevated price of salt and from the taxes that long ago were visited on this essential, but often rare, commodity. Besides such traces in the language, the gabelle induced reflection, from early on, about what constitutes good government.

AN ADMONISHMENT TO A KING. In 1699 at the end of his life, Vauban (1633–1707), marshal of France, published a book in which he offered a rational analysis of the various sorts of taxes levied in France and recommended the institution of a single tax, for purposes both of justice and of increasing royal revenues. He notes that: "The costliness of salt renders it so rare that it causes a kind of famine in the kingdom, felt most acutely among the common people, who, for lack of salt, cannot salt-cure meat for their use. No household can feed a pig, which they no longer do because they do not have what is needed to salt it. They do not even halfway salt their cooking pots, and often do not salt them at all."[40] Vauban lays out in six points his vigorous critique of the gabelle: (1) the saltworks do not belong to the king; (2) the saltworks are poorly guarded or not guarded at all; (3) and (4) numerous individuals and communities benefit from lease income on saltworks and sell the surplus; (5) regions exempt from the gabelle require the costly maintenance of armed forces stationed on their perimeters; (6) geographical variation in salt prices is the source of contraband trade.

The tragic nature of Vauban's text stems from the fact that it remained a dead letter. An aged Louis XIV did not follow up on it, except to order its seizure. Too many individual vested interests—an imperturbable Vauban bravely drew up a list of them, starting with the farmers-general—opposed rationalization of the tax system and elimination of the various privileges and abuses. It was only with the advent of the Convention, almost a century later, that the recommendations of the aging soldier would at last be ratified.

Vauban's tone is that of a man of experience, of fine sober judgment, having observed and noted what he saw over the course of "the errant life that [he] led for more than forty years." His brief volume inaugurated the fields of political economy and demography in France (he proposed, among other reforms, the establishment of regular census taking). The sustained indignation of its author, a great servant of the state throughout his entire life, lends the scent of a satirical tract to the work, which foretells the Revolution in its description of the failings of the ancien régime:

> It is certain that the King is the Political Leader of the State, as the Head is of the human Body: I think no one could doubt this truth. Moreover, it is not possible for the human Body to suffer

injury to its members without the head suffering from this. One can state that it is likewise the case with the Political Body, and that if the illness is not so rapidly conveyed to the Leader, this is because it shares its nature with gangrene, which, advancing bit by bit, does not neglect to encroach and to corrupt, along the way, every part of the body that it touches, until, having neared the heart, if it does not end by killing the body, it is certain that the body escapes death only through the loss of one of its members. . . . Kings have a real and very crucial interest in not overburdening their people to the point of depriving them of the necessities. (193)

This was a thoughtful and thought-provoking pamphlet. Vauban, whose figure would have nicely fit the Enlightenment, was way ahead of his time.

TAXATION.   I now expand the scope of the suggestion so as to draw the connection between the gabelle and taxation per se. By way of introduction to such financial matters, which as I will show amount to cold violence, let me note first that *salary*—both as a concept and as a word—originates with salt. As part of their remuneration, Roman legionnaires received their share of salt, or *salarium*, a word that then came to designate any regular payment, including cash (coins) rather than goods or services.

It is doubtless not by chance that the functionaries of war—that is, of institutionalized violence—were paid in salt. In reality and in principle, the sovereign mints money, and this money becomes legal tender in the territory over which he has authority. In tyrannical regimes, those in which power is based on violence, it is clear that violence is the foundation of money. In republican regimes, since national sovereignty is guaranteed by armed forces, violence also stands at the origin of money (if only through the insurrection that overthrew the preceding dictatorial regime).

Taxation may antedate money. In ancien régime France, most taxes could be collected in kind. And the central power, that of the kings of France, farmed out the administration of taxes, putting it up for tender to the highest bidder for time periods of varying lengths. This was the case, in particular, with the salt tax. Each of the gabelle salt warehouses had its farmer, that is, a person whose duty it was to levy and collect the tax, primarily for profit of the king and, to a lesser degree, for his own profit. This is how, starting in 1598, a single company, the Josse's family business, made itself the successful bidder for salt warehouses in the *généralités* of Paris, Châlons, Amiens, Soissons, Rouen, Caen, Orléans, Tours, Bourges, Moulins, and Dijon for a period of five years.[41]

In a nation such as ancien régime France, which was a mosaic of quasi-autonomous regions, regional delocalization of tax collection was doubtless a practical necessity. Therefore the convenient method of auctioning off tax collection to tax farmers was generalized. Farmers-general were the most powerful of the tax farmers; huge sums of money passed through their hands, from which they deducted the amounts due them—plus a bit. To have wed the daughter of one of these farmers-general was probably in the final analysis what cost Lavoisier his life during the Terror.

It is simple enough to draw comparisons with the farming of the gabelle. We may believe that the state enjoys today, and has done so since at least the American and French Revolutions, a monopoly on tax collection. In fact, this centralization is far from absolute. One can easily show that at the threshold of the second millennium certain specific tax farms continue to exist. As a first example, consider the collection of tolls on highways built and maintained by private corporations; some of these highways are highly profitable, especially after construction costs are completely paid off. The second example, less obvious and more subtle, is that of the drinking water supply to large urban population centers. It likewise is farmed out to private companies (such as the *Lyonnaise des eaux* or the *Compagnie générale des eaux*) at rates for consumers-taxpayers that, just like those of salt under the ancien régime, can vary by a factor of one to ten, depending on location.

Judging from the various forms of tax to which the French people are subjected today, the odious salt gabelle still endures in other forms of taxation. The most obvious—since, just like the former tax on salt, it affects an essential, regardless of social class and of resources—is the tax on energy.

In France as well as in most of Western Europe, a minute portion of the price of gas at the pump is collected by the refinery; the greater part is made up of taxes. The situation is the same for other taxes on energy (natural gas, fuel, electricity generated by various sorts of power stations, urban heating systems, etc.) paid by companies or by individuals. These are all gabelles.

In a logical extension, all indirect taxes, whether applied to so-called vices (tobacco, alcohol, PMU [PMU stands for *Pari mutuel urbain*, which is the name of the state-controlled system for betting on horse races.— Trans.], lotteries) or to other activities (public transportation fares, postal rates, and so on), are all preserved gabelles, whose injustice flows from the fact that they weigh more heavily on low-income households. The gabelles become all the more unfair when they are imposed on access to information, on self-teaching substitutes to education for families from disadvantaged sociocultural backgrounds unable to benefit fully from public education: radio and television taxes; taxes on newspapers and periodicals; taxation of telephone communications; taxes on books. One has reason to fear that governments, individually or collectively, might attempt to extend these gabelles to the Internet: this is one of the important stakes in the growth of its spectacular influence and success that we are presently witnessing.

Modern taxation suffers from the historic stigma of being heir to the taxing of salt. Even if, in a democratic regime, a tax is levied by the parliament to subsidize a public necessity such as education, public health, defense, highway infrastructure, and the like, it retains the indelible mark of its origins. Those origins were abuses of power. A lord demanded payment to grant the right of passage through his territory to mule convoys carrying salt shipments. Another lord, because he stored salt in depots for the public supply, received a fee not only in exchange for this service but because he had a monopoly on it. In both scenarios, taxation settled at a level that was just barely acceptable to consumers: making them pay more would have been counterproductive because it would have led to fraud on a grand scale or caused unrest. Our governments, which have to face considerable public debt, do not deny themselves comparable opportunities.

But one must be fair: taxation is not only a feature of national sovereignty, weighed down with vestiges of a feudal order. It is also a tool for the redistribution of wealth, serving the effort to achieve equality. And it can thus also be a control lever for the economy, increasing the relative value of the purchasing power of the disadvantaged, of those with lower incomes, through allocations (like, for example, the RMI [The RMI is a state minimum benefit payment to those in France with no other source of income.—Trans.]), deducted from the tax revenue allows the government to restimulate domestic consumption and thereby boost industrial production. Subtle adjustments to the income tax, and the interplay of direct and indirect taxation,[42] allow a state to foster—depending on the state of the economy—either individual savings or, on the contrary, increases in individual consumption. The feudal order, as symbolized by the salt tax, was also levied on the people slaving in the salt mines.

A MINE NEAR KRAKOW.   For centuries, salt was something of a curse. In czarist as in Stalinist Russia being sent to a salt mine, to the sadly renowned gulag, was the equivalent of being sentenced to Devil's Island. I once had the opportunity to visit an Eastern European salt mine, the most famous in Europe; I was left with dark memories of it, colored as they were by my introduction, the same day, to the world of the concentration camps.

In fact, it has become a ritual: a visitor to Krakow is shown the famous salt mine of Wieliczka, several kilometers southeast of Krakow, for a good half-day; then the tour bus takes the road to the equally famous camp of Auschwitz. A day of despair. I experienced this, twenty years ago or so. It's a region with a continental climate, summers of leaden heat and harsh winters. The locals are resigned to their fate: centuries of Eastern despotism have brought them to their knees.

I recall, of course, the salt mine. Since it had become a tourist spot, the guides had made it into a sort of cavern attraction: "See over there," they said, "we call that the Capuchin. This other block, that's the Young People." The attempt to make inert mineral one's own through naming and anthropomorphizing it relaxes, interests, and amuses. And the visit ends, of course, at the tiny chapel hewn out of salt by the miners.

One descends staircases with steps cut out of salt, one trudges along galleries hewn from salt, the chapel is decorated in statues carved in salt. The image comes to mind of ants wandering through a mineral maze in a network of dull sea-green passages, enclosed in ramparts of salt, beneath megatons of the mineral.

What sort of labor must this have been in this changeless and lifeless place? I picture the throng of workers amid the din of picks and shovels, assailing the blocks of raw salt, passively obeying orders shouted by the overseers in the pale glimmer of the oil lamp, for centuries on end.

One has the same feeling in the salt mine as one does at Auschwitz—Oświęcim in Polish, I believe—namely, an impression of the banality of evil (as Hannah Arendt wrote about Eichmann). These days, the camp is nothing more than a desert, one of those deserts that tourism makes of everything it touches. The banality of evil, when the slave sites of yesterday become the high places of today's tourism: Alcatraz, Devil's Island, the Isle of Gorée. The bite of frost. The bite of salt. The mantle of death. The gnawing of time.

THE WARRIOR'S SAYING.   Let's return to Aristotle and his notion that every concept should be looked at both ways, frontside and back. Let's switch viewpoints and move from the harsh injustice occasioned by salt to the justice and loyalty that, on the other hand, it can prompt.

The Japanese saying "taki ni shio o okuru" can be translated as "to conquer fairly." It immortalizes a historical episode that has become mythical: as a battle stretched on, a general sent over rock salt to the commander of the opposing forces.

"Enemy" is *taki*; *shio* is "salt"; *ni okuru*, an expression to which I will return, means "to send as a gift." Therefore the literal sense of the expression is: "to send salt as a gift to one's enemy."

The verb *okuru* has the primary sense of an "inferior's" greeting to a "superior": "friend" is *todomachi*, and "to hail a friend who is leaving" is *todomachi o okuru*. The idea of hailing is transformed into the idea of the gift one gives in friendship or civility: *mono* is an object, and *okurimono*, literally "object for hailing," means a gift.

This proverb recalls the chivalric virtues associated with the warrior or samurai. The hereditary aristocracy of samurais, maintained by the power of the shoguns, was enthroned under the Tokugawa. It comprised about 6 percent of the Japanese population at the time of our seventeenth century. Such institutionalization, reminiscent of the devising of the court at Versailles by Louis XIV, aimed at total control of the warring class by giving it a dependent, functionary status, preventing any unrest on its part by making it thoroughly subject to the central power that gave it its livelihood and controlled its resources and expenses.

The samurai ethic today undergirds the behavior of the Japanese in business. To take only one case, the revaluation of the yen under pressure from the Americans provides an example of "sending salt to the enemy"; it hardly slowed down Japanese success at exportation and, more generally, little diminished their ascendancy in the economic war.

CITADEL OF SALT.   But enough on the topic of chivalry. Back to the oppression of civilian populations by political powers. The Wieliczka salt mine, whose violence against its workers preceded but closely resembled the naked and barbarous violence of Auschwitz, is not, and by far, the only lasting manifestation of oppression. Another gave itself concrete expression in a monument of civil architecture, a very curious and exceptional masterpiece: a sadly uncompleted citadel of salt.

Claude-Nicholas Ledoux, born in 1736, was appointed chief administrator at the Lorraine and Franche-Comté saltworks in 1771. In 1774 this architect began work at the construction site of Arc-et-Senans, near Chaux; the project he was in charge of was to industrialize the production of igneous salt, burning wood—wood provided in abundance by the impressive forest of Chaux—as the energy source.

Although the saltworks at Arc-et-Senans remain unfinished, eleven buildings were constructed.[43] And Claude-Nicholas Ledoux left his design for the remainder in his notebooks and drawings. Today, a renowned cultural center has been established there.[44] The architecture is termed visionary, even utopian. But it is enough to call it prerevolutionary. I would call it grandiose (rather than functional, conceptual, well-adapted, ceremonial, or brutalist). A grandiose architecture aims to express sovereignty and the respect owed to power through the powerful clarity of simple shapes like the cube and the sphere; it rarely denies itself reference to the temples of antiquity, thus suggesting that power emanates from God or the gods and has a sacred nature. Certain elements, such as a triangular pediment and/or a colonnade, suggest this reference.

Claude-Nicholas Ledoux's saltworks belong to this type of architecture. It is in the neoclassical style. The eleven completed buildings consist of the director's residence; the two buildings for the *bernes* (or of the salts), east and west; the marshalry building; the guard building;

the two berniers,[45] the gabelle, the clerks' building, east and west; the coopers' building; and the stables.

The guiding concept here is a sort of Taylorization *avant la lettre*, or at least a corporatism: the saltworks represented the integration of various work processes but also of the various guilds associated in salt production. And each of these productive or social functions (for example, Ledoux had planned a brothel in the shape of a phallus) is both housed and represented in its own building. The architecture thus is both grandiose and symbolic. It prefigures other industrial architectural complexes, paternalistic and ostentatious, captivating by the coherence of their design, that of a self-sufficient whole, such as the Cité Mame in Tours, Considérant's Phalanstère, Godin's Familistère, and the Grand-Hornu, a mining complex near Jemmapes in Belgium.

The Arc-et-Senans saltworks, where salt was produced through brine concentration,[46] today resembles a concentration camp complex, perhaps in part because the construction work, having been stopped in 1779, has remained incomplete.

Ledoux's vision is best served, it seems to me, by the superb renovations recently performed on one of the Parisian city toll offices also built by him, near the Cité des sciences et des techniques (Center for Sciences and Technologies), which now opens on the view of a canal. This rescue from physical decay is also in a sense a restoration of an eighteenth-century political and civic culture whose triumphalist survival is demonstrated in today's United States. Many of the monuments in Washington, D.C., and many of its public buildings, on the Mall especially, obey such an aesthetic.

THE PROVERB OF THE CARDINAL POINTS.   The Irish proverb "go dtéighidh soir siar agus sac salainn air" means "until East goes to West carrying a sack of salt."[47] It is the equivalent of the French "when hens have teeth." A proverbial expression of this type in English is "when hell freezes over."

Most of these sayings, all paraphrases of "never," belong to the three categories of fantastical animals, geographic marvels, and calendrical anomalies. Besides the toothed hens, examples of the first category include winged pigs ("when pigs fly") and, in Rabelais, the arrival of the *coquecigrues*. [A term for a fantastical animal, something like a cross between a rooster or crane and a hemlock plant. "A la venue des coquecigrues" is an expression meaning "when the *coqueci-grues* arrive," that is, "never."—Trans.] The second category is illustrated in the proverb "when Calais and Dover meet." The third category includes the Greek calends ["Renvoyer aux calendes grecques" is an expression meaning "to postpone something indefinitely."—Trans.], the week of four Thursdays, and the Hispano–American *mañana*, meaning "it can always wait."

The quoted Irish proverb is more interesting for its excess meaning than for its belonging to the second category: it would have been enough to say "when East is mistaken for West" in order to convey the point. To add that the East, the Levant, will be the bearer of a load of salt (in other words, a gift of salt) attests to a whole people's resentment, no doubt of the English, on whom it was dependent for its salt supply.

After the oppression to which this Irish saying still bears witness, it is time to turn to liberation from the dictatorship of salt and, above all, from salt taxation.

GANDHI.　It was an immense beach. In the far distance, the eye had trouble distinguishing whether it was broken here and there by rows of trees or whether it ended at the horizon, before this greening. Likewise, it was difficult to make out the strand that extended, as far as the eye could see, in sand banks or in muddy stretches made of sun-hardened mire. The clay was coated with whitish patches, like an elephant's hide after bathing in a pond.

A curious ceremony was taking place at the water's edge in the early morning. A crowd, thousands of people dressed in white, was being led by a priest—or at least by a figure that a naïve observer might have fancied as such.

Without removing the round spectacles he wore, he advanced into the water, though his faithful followers stayed back on the beach, silent. The priest—but was he indeed one, or a mere preacher, some other sort of spiritual guide, a tribal chief, or even perhaps a physical education instructor?—entered the sea until he was knee-deep. He splashed it about his body and for a few instants made some ablutions, as if to partake of a purification ritual. The assembled crowd just stood there, in the back, attentive.

Then the man in white bent over the slack sea stood up, turned around, and resumed his walk, returning to the beach and to the group of his numerous followers. On the way back, he crouched over one of the sea's tidal leavings encrusted by the sun on an outcrop of dried mud and scooped up a handful of the white powder.

The crowd stood transfixed. At the very moment Gandhi picked up a bit of this coarse salt, in the total silence with which all of his actions had until that point been greeted, his gesture elicited a collective exclamation that resembled the roar of a wild animal and a sigh of relief.

This happened on the Kathiawar coast of India, at Dandi, on April 6, 1930. The scene took place in the early morning. The previous evening, the mahatma and the thousands of his followers who accompanied him had reached the end of a four-hundred-kilometer hike. At the age of sixty-one, Gandhi, followed by seventy-eight of the faithful on his departure from his Sabarmurti ashram,[48] had spent three weeks crossing the plains of Gujarat to the sea. His little troupe of marchers had grown bit by bit. In each village it went through, villagers would join the procession. And the local chieftain, appointed by the British vice-royal administration, had often resigned his post of his own initiative.

With this salt march, Gandhi had chosen to dramatize the demand for his country's independence. After long having planned it, he undertook the march on the advice of the poet Rabindranath Tagore, and only after having written a long letter, in a very friendly tone, to Lord Irwin, viceroy of India, in which he laid out the grievances of his people in a last conciliatory effort to avoid confrontation.

To the Indians, salt was symbolic of their colonial dependence. Rich or poor, everyone was subjected by the British Crown to a tax on salt. Like any other indirect tax, this particular one was unjust because it struck everyone regardless of need and income. Moreover, it applied to a commodity of the highest necessity, essential to the diet and to life. And it was quite a paradox—and an insult, too—given that India produced its own salt, to prevent all Indians from free and unregulated access to it.

For a long time, since 1919, Gandhi had mulled over this abuse of power and contemplated some protest action. Giving it the form of a salt march proved to be a cunning tactic. The colonial powers in New Delhi, caught unaware, were uncertain about what action to take and did not dare issue an order for the arrest of the mahatma. Gandhi's gesture, bending down to scoop up a handful of salt, set off a great explosion: a great many Indians went on to collect their own salt and set about supplying it to their compatriots, violating the law and slashing into the British Crown's monopoly. At last the authorities reacted with repressive force. More than sixty thousand Indians, having thus scoffed at the legislation put into place by the occupying British, were arrested. On May 5, 1930, Gandhi himself was apprehended as he slept peacefully by the sea under a mango tree.

On May 21, 1930, Sarojini Naidu led a group of two thousand nonviolent protesters to the Dharasana salt production plant, which was surrounded by police forces. As each column of marchers met the ranks of police, it was attacked with blows from *lathis*, special whips reinforced at the tips with steel. The protesters responded to the violence passively, enduring it. And another group of marchers would take the place of the group before it. At eleven in the morning, the demonstration was over: two dead and three hundred twenty wounded were taken away in the scorching heat.

The Salt March stands out from previous uprisings provoked by salt taxes such as the gabelle whose harshness made them unbearable. The civil disobedience launched by Gandhi was nonviolent, in a context of fidelity to traditional social and religious values. Gandhi invoked the *Bhagavad-Gita* and prayed to the gods of the Hindu pantheon to guide his actions, above all to help him maintain his moral purity.

In conceiving of the Salt March, Gandhi put into place a new social contract. The previous one was null and void, since England had imposed it on India by force: in the words of Gandhi, India consisted of "three hundred million people, oppressed and living in terror of three hundred men." India could and should stand up for its autonomy and rely on its own resources alone; this is also the message of the Salt March. Moreover, a social contract is built on popular sovereignty: the people must be sovereign in order to delegate its sovereignty to its representatives. The Salt March expresses and makes manifest such sovereignty: in the final analysis, the Indian people indeed own their nation's territory together with its mineral riches, such as salt. "The nation," wrote Gandhi, "wants to gain an awareness of its strength even more than it wants to obtain independence." For this reason, it was incumbent on the reunited Indian people, joined in solidarity, to provide for themselves both a political system and representation. It would do so, in spite of its fragmentation into different ethnicities and religions, once it had reasserted its ownership of the land.

As the Boston Tea Party symbolized the American revolt against English power, so this Salt March stood for the uprising of the Indian people against the same British power and for similar reasons: an arbitrary and overly burdensome tax on a commodity of the highest necessity.

The Salt March was one of the great moments of the twentieth century. It struck one of the first blows against the colonial English occupiers in the struggle for Indian independence waged by the Congress Party; that independence would be acquired definitively only around fifteen years later, in the aftermath of the Second World War.

The march served as a model for other protests against exploitation or oppressive colonial power. The nationalization of the Suez Canal in 1956 by Gamal Abdel Nasser can be viewed as one of its direct descendants. Collective memory has likewise not forgotten the marches for the equal civil rights of black Americans led by Martin Luther King Jr. in the pacifist, nonviolent spirit of Gandhi.

With the Salt March, Gandhi took as his point of departure the sense of dispossession imposed by colonization. Colonized Indians had been made dependent on the English colonists. For generations, the colonial yoke had been so thoroughly fixed in place that the Indian was infantilized by his inferiority, which he internalized and felt powerless to change. He experienced as natural the course of things imposed on him. The colonizer, as a rule, imposes all his values: he substitutes his own culture for the culture of the colonized. The latter is weakened and finds itself in the grip of a more or less insidious force of destruction.

Gandhi's reaction, in light of this analysis, was to declare the exemplary nature of the need for table salt. He reasoned that when Indians became aware of their ability to get their own salt—at the level of each household and each village—this simple recognition would lead them to free themselves from domination by the English. Granted, it was something of a truism but a revolutionary one: revolt need not be violent; a war of independence was not unavoidable. All that Indians needed was to rediscover their economic autonomy in order for the colonial power, ipso facto, to show itself in its true colors: as a foreign body, an infection that the Indian nation could eliminate once it regained awareness of its own strength, a strength stemming from its resolve to count on no one but itself. It had to secure its autonomy, expressed in this case as self-sufficiency. Any Indian could boil some brine in a pot and in this way collect the salt needed for everyday life.

At issue here is a shift from the cultural sphere to the technological sphere. Gandhi had moved from political analysis to defining both the leverage (the large Indian population) and the fulcrum (from now on, each Indian would make his own salt and thus satisfy an inescapable daily physiological requirement). To the Indians, colonization had obscured their own technical knowledge: every Indian was in fact capable of providing his own salt, as timeless knowledge had allowed him to do before colonization. And if the Indians were once again capable of fulfilling their own needs for salt, they could free themselves from British domination.

Generalization was the crucial point. Salt is not a special case: all other vital needs of the Indians could be satisfied by recourse to local, preindustrial technologies. This is why Gandhi chose the spinning wheel as the symbol of the nonviolent struggle for independence. India did not need to import textiles made by British industry. It would boycott even those products it was forced to buy in favor of substitutes it would manufacture once again.

Gandhi's argument forms a whole, and it would be a mistake to limit it solely to its political and oppositional dimension. The nonviolent actions he chose all share the distinctive characteristic of formulating the struggle for collective rejection of colonial power as the sum of an infinite number of individual actions. Furthermore, these actions relied on soft technologies, as we would call them today: they restored ties with tradition, and—as their supreme virtue—they were ethical, for they let each Indian take root once again in an authentic way of life.

Gandhi thus worked out an extraordinarily effective model of political struggle, one that has been influential throughout the twentieth century. It is indeed worthy of admiration but also bears within it the seed of various pathologies: to begin with, a certain angelic, excessive attachment to the past, the danger of which I will later review, whether in the Chinese Cultural Revolution—for example, with the artisanal blast furnaces aimed at increasing steel production, which proved to be a colossal failure—or in the Khmer Rouge seizure of power, which forced an exodus of city dwellers to the country in the name of the creation of a New Man.

five **biology**

The ocean is salty, and its saltiness keeps increasing with the bite of time: rainwater dissolves and sweeps away mineral salts from soils, and rivers then carry them to the sea. Conversely, water evaporation, a distilling, pumps pure water into the atmosphere. In this way, salinity gradually increases.

Life on Earth goes back to marine organisms, some of which still remain relatively primitive. The earliest forms, single-celled beings, both differentiated and shielded themselves from the marine medium with a membrane. Reproduction into an identical cell has tended to preserve within each cell the composition of the primeval ocean, hundreds of millions, if not billions, of years old and less rich in salts than the present-day ocean.

Thus the cell membrane is not only the cell consciousness, preserving the memory of long bygone eras. It also sieves the chemicals that pass through it; it has to maintain the integrity of the intracellular environment, which for marine organisms, as I just noted, is less salty than the external environment.

In fact, there are two kinds of organisms: those whose intracellular contents adapt to changes in the outer salinity (osmoconformists) and those able to maintain their internal composition invariant in the face of salinity fluctuations (osmoregulators). Jellyfish and crabs live in environments of varying salinity and belong to the first category. Fish belong to the second. Freshwater fish and saltwater fish thus need differing adaptation mechanisms. A species such as the salmon, shuttling between both sorts of waters, has evolved for itself sophisticated means for acclimation.

In human beings, the relation of salt to thirst, which intuitively we may believe to be simple and elementary, actually attests to a complex physiology in which the kidneys and the brain both play a role. In addition, organisms take advantage of the differences in salinity between intra- and extracellular environments for other ends, one of which is the nerve impulse.

I will then discuss the strange case of extreme halophiles, that is, single-celled organisms that manage to live in environments with very high salt concentrations. And I will quote, in closing, a description of the Great Salt Lake by an early French traveler writing in the mid-nineteenth century.

THE SALINITY OF THE OCEAN.   How did the sea become salty? We have many ideas, but we don't know. Answers to this question vary enough to indicate our profound ignorance.[1] Furthermore, such speculation on origins is metaphysical. It belongs with historical studies and can give rise only to the devising of more or less plausible scenarios and accounts. In this respect, we have not divorced ourselves from scholasticism. On this topic of the origin of the ocean's salinity, scholasticism, at least till the Renaissance, contented itself with repeating and discussing Aristotle's answer.[2] Aristotle began by refuting Empedocles, for whom the sea was the sweat of the earth. For Aristotle, the sun draws off fresh water and in so doing leaves behind a brine.[3] Such brine obtains because the dry exhalation issued from the land mixes with the vaporous exhalation from the ocean, and when this mixture falls back down as rain, it supposedly makes the sea salty.

The very existence of oceans on Earth itself remains a mystery. It is believed that the early earth had numerous collisions with comets and received from these continual enrichment from the comets' supply of water (the tail of a comet is made of dust coated with dirty ice). Would such accretion ultimately result from the hydrogenation of carbon dioxide (CO_2),[4] yielding water together with the carbon-containing carbon monoxide (CO) and methane (CH_4)? Another theory proposes instead that oceans derive from the decomposition of silicate rocks (such as present-day clays) and the resulting water loss.

The contents of present-day oceans average 47 millimoles of sodium and 546 millimoles of chloride per liter (the mole is the unit of mass used by chemists).[5] The primeval ocean would have been less rich in salt(s) by a factor of about three.[6] Salt came to the ocean from erosion of the earth's crust by rivers,[7] and evaporation of water might increase its concentration, but evaporation, followed by condensation into rain, involves only pure water.[8] So the abundance of salt in the seas has to reflect the abundance of the elements sodium and chloride in the earth's crust, which in turn would merely attest to the high abundance of these two elements in the cosmos as a whole.[9]

A MARINE ORIGIN?   Even though the hypothesis—one I cannot vouch for—that life began in the primeval ocean or in tidepools (Aleksandr Ivanovich Oparine's hypothesis of the "primordial soup") is attractive, the evolution of species did move from an aqueous environment, that of fish, to the air environment of the higher vertebrates, the class that includes most mammals.

Consider first the simplest, as well as the oldest, case from an evolutionary standpoint, that of single-celled organisms. They embody compartmentalization; a membrane separates their interior from the outside world. This membrane both separates and joins: each cell must have continued exchanges (such as nutrition and excretion) with the outside environment. Various chemical species thus cross the membrane border in both directions.

Reproduction is the process by which species achieve a form of permanence, while their individual members die. Since ocean salinity did increase over the course of time, a given organism had either to adapt to this change and gradually enrich itself in salt or find a way to maintain the integrity of its intracellular salt content. As a matter of fact, many kinds of organisms, single as well as many celled, show salt concentrations generally lower than that of present-day oceans (about 35 grams per liter), as if they had preserved in their cells the memory of the composition of the primeval ocean. But how did they manage to maintain such integrity while facing a gradually saltier environment?

WHAT OSMOSIS CONSISTS OF. We have all watched bubbles. An air bubble rises in the water, as at the start of the Jacques-Yves Cousteau film *Le monde du silence* (*Silent World*). As it rises, gradually its volume increases. The external pressure—from the weight of the column of water on top of the bubble—slowly lets up, and this allows the air inside the bubble to expand. The same is true for a balloon filled with a light gas such as helium (or with hot air): the higher it climbs in the atmosphere, the more its contents swell because the outer pressure—that is, the weight of the column of air on top of it—steadily decreases. The balloon's skin is the surface of separation.

A bubble's or balloon's spherical shape and its volume balance two pressures: the inner and the outer. The bubble and the balloon remain still when the two pressures are in equilibrium; otherwise, they move in the direction that will establish or reestablish this equilibrium. When a balloon's skin is pierced, the deflation can be catastrophic. This is what one observes with a soap bubble, that is, with a small amount of air enclosed in a fine membrane of soapy water: this iridescent, thinnest of membranes, if pierced by running into a sharp obstacle, deflates suddenly and vanishes.

Biological cells are just like little sacks, too, spheroids at first sight; a casing separates their interior from the exterior environment. Both these environments, the intracellular and the extracellular, are made of salt water containing various other chemicals.

The intracellular compartment accounts for two-thirds of the water in our bodies. The other third is made up of blood, which carries blood corpuscles (these are also cells) throughout the body, and of the water supplying the cells in the various tissues, muscles, kidneys, or the brain.

Say you eat some salty snacks: pretzels, chips, or a piece of dry sausage. The amount of salt (sodium chloride) in extracellular water increases almost immediately. From then on, the extracellular environment is saltier than the intracellular environment. Faced with this disequilibrium, nature has perfected a mechanism for the selective permeability of biological membranes, or *osmosis*. Water escapes cells through their membranes, to rejoin the extracellular sodium, but at a cost, namely, that of the dehydration of the cells. So, you have to drink to return the volume of intracellular water to its previous level.

Keep in mind this simple idea: osmosis refers here to entrances and exits, across the membrane border, that are allowed to water but not to sodium.

THE TWO KINDS OF ORGANISMS.   But let's return to those beings that live in or near the ocean. They belong to at least two classes, depending on whether they adapt to the surrounding salinity and its variation (these are named osmoconformists) or, on the contrary, whether they control the salinity of their internal fluids (known therefore as osmoregulators). In the first class are jellyfish, certain mollusks, and some crabs.

A simple experiment allows one to determine whether a marine animal is an osmoconformist or not: one has merely to weigh it before and after immersion in fresh water. If it is an osmoconformist, water enters its cells to decrease the disproportion between the internal and external salt contents of the tissues, causing an almost immediate increase in weight. One can easily demonstrate this with live mussels prevented from snapping shut by a small object such as a coin or a paperclip placed between the two valves of their shell. Immersed in fresh water, they show a distinct increase in weight, whereas a control specimen from the same group of mussels, one allowed to close up, remains the same weight.

Crabs include representatives of both categories. Some crabs, in habitats such as estuaries, where salinity varies depending on tides and seasonal changes in humidity and aridity, are osmoregulators. Placed in less salty water, the crab eliminates a large quantity of salt in its urine. Conversely, its branchia allow it to take in a full supply of salt again, if necessary. Certain species, such as the hermit crab, or *Clibanarius vittatus*, are capable of osmoregulating in conditions of low salinity and swing to osmoconformity at elevated salinity levels.

FISH.   These vertebrates shed salt in their urine and absorb salt through their gills. The cells of a freshwater fish have a salinity greater than that of its surroundings. Therefore the fish loses salt to the surrounding environment and, conversely, receives an influx of water. To counteract this double threat, the fish produces large quantities of urine, dilute in salt, so as to lose water. Its gills actively absorb salt, that is, sodium and chloride ions, to make up the losses.

Conversely, a saltwater fish has gills that are nearly impermeable to water and excrete sodium and chloride ions. Moreover, the kidneys of a saltwater fish excrete a small amount of urine (to minimize water loss) that is highly concentrated in salt (to expel the most salt).

A fairly large number of fish species live in both environments: these include lamprey, sturgeon, shark, trout, and salmon. Most have adapted to changes in salinity. Among the Salmonidae, which include salmon, the fish even undergoes a number of anatomical and physio-logical changes controlled by signals such as the changes in tempera-ture and sunlight that accompany the return of spring, an entire moulting that prepares it to cope better with its new environment.

THIRST AND LACK OF SALT.   Might one define animals as living creatures with a craving for salt? Animals indeed have a universal phys-iological need for salt. In order to obtain this ingredient essential to their survival, some of them achieve levels of activity worthy of the *Guinness Book of World Records*. Witness the behavior of the *Gluphisia septentrionalis* moth, recorded by Cornell biologists Scott R. Medley and Thomas Eisner.[10] A centimeter and a half long, the male of the species displays apparently obsessive potomaniacal behavior: it spends hours (up to three and a half at a stretch) drinking in puddles of water. Since it would thereby absorb as much as six hundred times its own body weight, it certainly needs to expel this water once it has obtained from it what it seeks, that is, sodium in the form of salt. The insect thus makes sure to emit the excess fluid, depleted in sodium, beyond the puddle from which it drinks: every three to five seconds, his rectum expels a powerful stream more than fifty centimeters long.

Why such activity, why such a need to hoard sodium salts? The male stockpiles this reserve store before presenting it to his mate in the form of sperm, during a coupling that lasts about five hours. The female in turn will transmit most of the sodium thus delivered, in the form of a sperm packet, to her eggs, that is, to the young *Gluphisia* to come.

Human beings are no exception to this pattern: salt deficiency induces a feeling of need and a compulsion to satisfy it. This innate need is not cultural. Moreover, it contrasts with thirst in that it is not immediately perceived, which can become dangerous: experiments on sheep show a lag of two to four days between the objective need for salt and its satisfaction. Mammals also have mechanisms for salt retention tied to the secretion of hormones like aldosterone.[11]

What about thirst? Eating salty foods makes one thirsty.[12] The phenomenon can be traced to the basic physicochemistry of water passing through cellular membranes so as to achieve the equalization of salt concentrations on both sides of the membrane.

Thus water is more or less toxic depending on its lack or excess of salt. When it contains too little salt, its ingestion tends to increase the volume of cells that come into contact with it (turgescence). Accordingly, blood serum tends to become too dilute in salt and thus weakened in osmotic pressure. This is why thirsty mountain climbers are advised not to consume snow or drink melted snow.

Seawater is unsuitable for drinking because it contains too much salt. If one drinks it, the internal environment tends to become too rich in salt, which produces hypertension in the strict sense of the term; the internal environment is then known as hypertonic. This occasions a loss of water from the cells, which try quasi-mechanically to correct the state of disproportion that exists—by about a factor of three—between the extra- and intracellular saline concentrations.

Yet in the 1950s the French physician Alain Bombard very courageously proved that shipwreck was not an automatic death sentence.[13] The lasting component of his proof was his reliance on the inflatable pneumatic dinghy, which has now become part of the standard safety equipment on ships and airliners. But he also proved, contrary to customary opinion, that ingesting small quantities of seawater can help survival.

In vertebrates, the kidneys are the main organ for salt excretion, through the production of urine: the blood passes through the kidney, entering through an artery and leaving through a vein, while urine flows out of the kidney through the urethra. The filtering of blood is carried out by specialized units in the kidneys. These organs produce urine of variable concentration in order to maintain a constant level of intracellular salt concentration. In the cells of other organs, specialized and poorly understood "antennae," called osmoreceptors, measure this level. Changes in the level trigger neuronal signals to the brain, causing the brain to produce hormones, that is, chemical messengers. For example, the brain emits vasopressin, a hormone that orders the kidneys to stop expelling water, so as to avoid dehydration. In the opposite direction, the kidneys send another messenger, angiotensin, to the brain; it causes a sensation of thirst. Numerous other complex hormonal and nervous system controls govern these exchanges of water and salt.

THE NERVE IMPULSE.   Indeed, a change in the salt concentration translates into an electric signal within the nervous system. How does this occur? What is the mechanism for such a phenomenon, reminiscent of the battery invented by Alessandro Volta? In order to understand it, one must go just a bit deeper into detail. The whole process is predicated on the ability of cells in organisms to discriminate between the two closely related elements, sodium and potassium.

Animal cells expel sodium Na^+ and allow potassium K^+ to enter. Accordingly, these elements differ quite a bit in concentration between the inside and the outside of a cell: the concentrations in each of these ions differ by a factor ten. Besides such a concentration differential, cells make use of a chemical engine to pump in and out these potassium and sodium ions, in a ratio of three sodium to two potassium:[14] this amounts to the cell lining itself with electric charges, negative inside the cell membrane and positive outside. This makes it possible for a cell to generate an electric current: any live cell maintains itself with a potential difference, just as the supply of electricity to our households in the United States is at 110 volts of an electric potential difference. (Let's shorten it for brevity's sake to "electric potential.") While the energy supply to our electric radiators is at 110 volts and that of a Walkman just a few volts (often 6 volts), that of a biological cell is at an electric potential of several tens of *milli*volts (mV).

Such is the case for nerve cells, or neurons: at rest, they show an electric potential of about −50 mV. When a neuron is fired, the electric potential increases very quickly, moves through the value zero, climbs to about +50 mV, and then drops back to its rest value—in fact, going through an even lower minimum of about −60 mV. The surge to the maximum, known as the action potential, is due to the sodium ions bursting into the axon; initial concentrations at a factor of ten account for a change of +58 mV. This change is accompanied by the expulsion of potassium ions, which accounts for the drop back down to a minimum prior to returning to the rest value. The entry and exit of ions are represented in figure 2.[15]

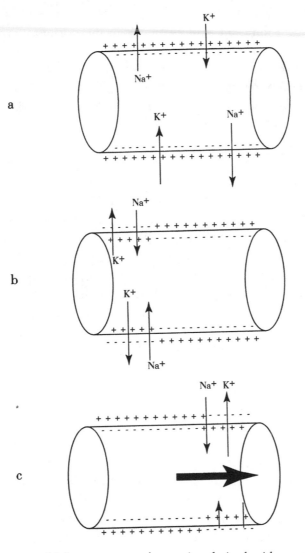

FIGURE 2. (a) At rest, a neuron's axon is polarized, with negative charges on the inside and positive charges on the outside of the membrane; for this reason, the membrane's electric potential is −50 mV. (b) Entry of sodium cations and exit of potassium cations, or depolarization, makes this electric potential increase to +40 mV before returning to −50 mV. (c) The nerve impulse is the movement of this depolarization along the length of the neuron.

The functioning of the human nervous system thus requires the maintenance of a sufficient amount of salt in the organism; in this respect it is just like all other cells in the body. This is why we have to absorb one to two grams of salt per day. A salt deficiency, stemming in general from overexcretion (we excrete 90 to 95 percent of ingested salt in the form of urine and sweat), is felt immediately: particularly in nocturnal muscular cramps in the calves and feet, as I know from painful experience.

The sodium Na^+ ions, with weak electrical charges compared to other metallic cations (calcium, magnesium) that serve as their physiological rivals, are extremely nimble, doffing or donning their watery coats in a flash. This ability to pass through barriers, which they share with potassium ions, makes them excellent messengers: this is why nature has chosen them as carriers of the nerve impulse.

In my laboratory, when we used nuclear magnetic resonance (NMR) to study the binding and the release of the sodium ion by biomolecules—in order better to understand the permeability of cell membranes, for instance—I believe that I innocently endowed this corpuscle with my childhood admiration for masked and agile heroes like Lagardère in Paul Féval's *Le Bossu*. The sodium cation takes after the lighthorseman or the lone commando!

EXTREME HALOPHILES.   Our present knowledge divides living beings into three categories:[16] the prokaryotes (bacteria whose cells have no nuclei), the eukaryotes (bacteria and all higher organisms whose cells have nuclei), and archaebacteria (primitive bacterial organisms). The last are distinguished by their mitochondrial RNA sequences, by the chemical composition of the cell walls and membrane—the membrane does not consist of a lipid bilayer but of glycerol bound by *ether* (and not ester) bonds to long chains of twenty or forty carbons—and by the presence of RNA-polymerases whose degree of complexity is between those of the prokaryotes and eukaryotes. Archaebacteria are not rare organisms: 30 percent of the oceans' pelagic organisms fall into this category.

Among these microorganisms, there are bacteria that live in extreme environments such as the hot springs on midoceanic mountain ridges. Another class of archaebacteria, the extreme halophiles, thrives in areas with high salinity that one might have a priori believed to rule out any life form: lakes or seas with salinity as high as 300 grams of salt per liter. For example, the African lakes, Nakuru[17] and Simbi, are so full of sodium that their pH reaches 10. Lake Nakuru contains 8 percent sodium. Millions of flamingos live on its shores and feed on bacteria such as *Spirulina*.

More generally, one finds extreme halophiles in salt marshes, evaporation pools rich in salt, hypersalty seas like the Dead Sea or the Aral Sea, subterranean salt deposits, salt domes, brines and salt curings, dried meats and fish. Bacteria such as *Dunaniella salina* live in aqueous solutions that can reach sodium chloride saturation, that is, the water is so laden with salt that it can no longer dissolve any more of it.[18]

One can only admire the adaptive mechanisms that allow these bacteria to thrive in areas a priori unsuited to life, with salt concentrations of between two and five moles per liter. The two chief problems to be solved by extreme halophiles (such is their name) are the considerable osmotic pressure on the outside of the cell and the precipitation of intracellular proteins through the process of salting out.[19] One response to the first constraint is that employed by *Dunaniella salina*, which produces elevated intracellular concentrations of glycerol (glycerin) in order to help balance the huge outer osmotic pressure.[20] Another countermove is to expel sodium and let potassium enter as an intracellular cation.[21] As a response to the second constraint, bacterial proteins are modified and made more hydrophilic by the introduction of acid residues—in particular, glutamic and aspartic acids—to their surface; some of these acid residues form salt bridges to the base residues lysine and arginine, providing greater rigidity and structural stability to these protein molecules.

That life can adapt itself to such harsh and seemingly inhospitable environments prompts us to greater intellectual openness and tolerance: we will certainly discover other forms of life in other environments that seem a priori out of bounds (why not under very high pressures, for instance?). Genetic engineering technologies will modify existing organisms to make them more able to survive in inhospitable places. Will humankind one day be capable of constructing artificial life forms that thrive in surroundings unimaginable today?

I deem it important to think about what is excessive and beyond the norm: scientific thinking has such a playful and provocative turn of mind. Its role includes the imagining of monsters; in this way, the science community resembles other social groups whose solidarity is also founded on the occasional orgiastic or carnivalesque excess: making merry and thinking out the limits of one's own knowledge stem from one and the same dionysiac impulse. Occasionally, the excessive is felt from a simple look at an unfamiliar scene or a bleak landscape.

A FRENCHMAN'S LOOK AT THE GREAT SALT LAKE.   When a fresh eye surveys a bleak and incomprehensible sight, nature at its most revealing and nevertheless also at its most bewildering, does such a puzzled look invite the arid quest for knowledge? At the end of the 1850s the French traveler Jules Henry, on his way from California, reached the new territory of Utah. After returning to Paris, he published in 1860 his *Voyage au pays des Mormons*.[22]

In spite his having the objectivity of the science writer, his description of the Great Salt Lake, is that of a desolate landscape, punctuated with subdued intimations of death:

The lake is a genuine Mediterranean sea, without any communication with the ocean. It is no less than a hundred leagues in circumference, and it must have taken up, in earlier centuries, a much wider area, since the geological aspect of the soil on our trail makes us believe that its ramifications extended deep into the Utah valleys. The existence of the Salt Lake had been suspected since 1689, as one can see from the memoirs of Baron le Houtan. Yet it is only in recent years that its position has become known definitely. When Mr. de Humboldt was visiting Mexico, the Salt Lake was still rather mythical, and the illustrious traveler proposed quite a few hypotheses. It lies between 40 and 42 north latitude and 114 and 116 west longitude. To the northwest its waters extend so far that the eye, unable to discern the mountains flanking it, is lured into believing that they extend indefinitely, like a vast sea. Its depth is inconsiderable: it does not exceed ten meters and is on the average seven or eight feet. In the middle of the lake, several sizeable islands rise up to one thousand meters and more above the water level. At present one does not see the slightest boat going through desert this sea. However, the aborigines tell of a tradition: the Utah Indians in former times would put large dugouts on it. Water is of such density as to prevent the human body from sinking. During our stay with the Mormons, we went and bathed in it several times; we would lie down on the surface, and we could stay there indefinitely without the slightest effort or movement. One might fall asleep without risk of drowning, it would seem. Such extraordinary water density explains why animals are unable to live in the lake. One finds neither fish nor mollusk there. The only representative from living nature to have been reported, although rarely, is a little worm in the beach sand. Trout, which sometimes make their way down to the lake on streams, die immediately on reaching it. The plant kingdom is represented by a single algae, from the Nostochineae tribe. The lake shore, in the north especially, is covered with a considerable layer of the most handsome salt, which is exploited for the needs of the people. When we were there, one could see on the shore and on top of the salt deposit, a layer one foot deep consisting entirely of dead crickets. Those insects, which a violent wind had carried away in the form of enormously thick clouds, had drowned in the lake. They had earlier, during the summer, destroyed the sowings and even the grass in the meadows. A food shortage ensued, and the Mormons saw in such a scourge only further vindication for the truth of their religion, because it had happened, just as with the Israelites, on the seventh year of their settling in this land.

The lake has no tides, but under the variable force of the winds, the water surface wrinkles, and small waves leave foamy spume on the shore. No trees are to be seen next to the lake nor in the nearby plains. One has to climb to the top of the nearby mountains to find wood for burning, such as green trees, some *Acer* [maple], *Salix* [willow], *Populus* [aspen], and *Quercus* [oak]. Next to the beach can be seen only a few dried-up plants, such as yellow Compositae, a yellow *oenethera*, and mostly a tall Cleoma with pink flowers.

Jules Henry then proceeds with a lucid analysis. He attributes the lake's extreme salinity to its recession and to evaporation, as some of the freshwater inflow had dried up in more or less recent times.

SEA SALT STACKS

© Kevin Schafer/CORBIS.

WORKERS SALTING THE HERRING, ca. 1943

© The Mariners' Museum/CORBIS.

RAIL TRANSPORT AT A SALT MINE, YEMEN, 1926

© Hulton–Deutsch Collection/CORBIS.

WOMEN COLLECT ROCK SALT, SRI LANKA, ca. 1935

DIRECTOR'S HOUSE OF SALINE ROYALE, ARC-ET-SENANS, FRANCE

Photograph by Georges Fessy © Institut Claude-Nicolas Ledoux.

MAHATMA GANDHI WALKING TO SHORELINE, 1930

Photograph © Bettman Archive/CORBIS.

FLOATING IN THE DEAD SEA

© The Purcell Team/CORBIS.

SALIERA [SALTCELLAR]

Salzfass (1540–43) by Benvenuto Cellini (1500–1572), courtesy of
the Kunthistorisches Museum, Vienna.

THE MORTON SALT "WHEN IT RAINS IT POURS" GIRL
FROM 1914 TO 1968

six **other science insights**

To borrow from French humorists such as Alphonse Allais or Pierre Dac, processing seawater for its salt or water would seem, logically, to have exhausted the subject. Since the days of the alchemists, however, water and salt have had still other joint projects. Electrolysis, the decomposition of salt water by passing an electric current through it, supplies the chemical industry with its sources of chlorine and soda. Adding salt to water—two substances whose mutual affinity, though mundane, nevertheless raises questions for scientists—also allows water to be cooled, a temperature reduction that can be attributed to the increase in disorder that occurs when crystalline salt is dissolved in pure water: if desalinating salt water requires energy, the production of salt water, on the other hand, produces entropy, or the energy of disorder, usable for cooling, which has been achieved in this way since 1620.

A glass of wine tipped over on a tablecloth: this everyday incident is the starting point for a presentation of chromatography, a valuable set of tech-

niques for separating and purifying substances. I will also explain why water is softened when it passes through a salt-containing material and why stoneware pottery gets a fine glaze from vaporizing salt at a high temperature in a kiln.

These scientific remarks are rounded off by noting other circumstances in which salt catalyzed creativity, whether the invention of spectroscopy by Bunsen and Kirchhoff or Zeeman's discovery of the influence of a magnetic field on spectral lines.

But the chief contribution of salt to science is without a doubt the notion of the saugrenu: *an unexpected event or phenomenon that exploded accepted knowledge and compels us to rethink the world.*

I close with reflections on the teaching of introductory chemistry, since it is altogether saugrenu *(in a pejorative sense) precisely when it deals with salt and more generally with salts.*

ALCHEMY.   For a long time, alchemy remained dualistic: matter was conceived of as issuing from Aristotle's four elements (water, earth, air, and fire) and as the product of strife or harmony between the two complementary principles, sulfur and mercury. Paracelsus, to whom we owe psychosomatic medicine and the beginnings of chemotherapy, introduced salt as the third alchemical principle. But which salt? As with sulfur and mercury, it was not a matter of ordinary salt but of the salt of the philosophers.

But here, precisely, is a description of ordinary salt, which I take from the *Dictionnaire hermétique*;[1] I give it in full, with my commentary interspersed.

"SEA SALT: This salt is composed of a great deal of Mercury or internal humidity for the fusion of a bit of volatile, combustible, and salineous sulfur and a good amount of the dry, or pure, earth, for its fixity united in their principles." Since the alchemical enterprise consists, on the one hand, of fixing the volatile and, on the other hand, of volatilizing the fixed, this text defines sea salt as a compound of the moist and the dry, of the fixed and the volatile. The moist would be the mercury principle, the dry would be the element earth, and the combination of a little mercury and of sulfur in a greater amount makes the latter dissolve, before making this product undergo another combination, this time with earth.

"Its very difficult melting makes manifest to us its internally cold nature." Sea salt is extremely refractory, that is, infusible: it has a cold soul—making it akin to ice—that opposes it to mercury's heat, as well as to sulfur's flammability.

"Its spirits are white." This is a reference to salt spirits, or hydrochloric acid, a colorless gas.

"And if it is acrid, a desiccant, and therefore dry and hot, this is only by chance, a result of the volatile salt and the combustible sulfur, its opposites, with which it is joined." Here alchemical thinking seeks to reconcile these apparently opposed empirical properties—salt's resistance to melting and its easy ability to produce hydrochloric gas in the presence of an acid—hence the repeated recourse to the qualities of the dry and the moist, the hot and the cold, qualities directly linked to Aristotle's four elements.

Now the text switches direction: after having set out orthodox alchemical doctrine, which our present understanding no longer grasps effortlessly, it presents reasoned, scientific arguments based on empirical observation of natural facts: "Some persons professing to have Science say that the sea gets its saltiness nowhere but from salt, from the very earth that is its womb, as the water is its wet nurse, since one finds some maritime beaches that are saltier than others and because one comes across various springs—seemingly salty ones—very far from the sea, deriving their bitterness from the earth itself and from ammonia." The allusion is to the arguments, grounded in clear, shrewd observations, offered by figures such as Bernard Palissy.

Alchemical thinking—which is symbolic and syncretic, certainly not scientific in our current sense—far from distancing itself from these arguments will very honestly continue to set them out: "Others say that it is only the rays of the sun that make the sea salty; and to the degree that the Sun shoots forth its rays more intensely over the sea's waters, the water there is saltier, and that where it shoots them forth less intensely, the water is less salty; and that all the other salts that are found in the three Kingdoms of nature have their origin in sea salt."

The author now comes to another series of relevant observations that tend to link evaporation of the sea to salinity through the heat of the sun's rays: "They hold further that when the salt sea waters flow out to make various springs and rivers, they travel through the ports [pores], that is through several small channels and seams in the earth, where they are filtered and leave behind their saltiness; that is why they come out as fresh water. This saltiness then is used by nature to produce various things, about which the Reader can engage in some fine and odd reflections."

My reflection, by way of a conclusion to this odd little hybrid text, is that it is still part of an organicist vision of the cosmos; the earth is likened to a giant organism, exuding salty water but also absorbing sea waters through all its pores so as to filter them and make them into fresh water.

MICHIGAN SALT.   As I have just noted, in one of the oldest metaphors, which continues to resurface at irregular intervals, Earth is viewed as a giant organism. This huge body sweats. It has both secretions and excretions. The visitor to, say, Yellowstone National Park is struck by these sight and smells: mineral springs, deposits of various salts, of pristine sulfur, jets of steam issuing from witches' cauldrons, the whole gamut of strange and vaguely threatening metabolic activities.

Native Americans integrated such manifestations into both their worldview and their livelihood. They knew of salt springs, which dot the North American continent, and they used them for their supply of salt.

There are many such natural sources, scattered across the United States. Michigan's Lower Peninsula, for example, boasts numerous effusions of salt brine. When European immigrants started moving in during the nineteenth century, they started the two associated industries of salt production and lumbering. Burning the trees supplied the heat for water evaporation in large kettles and thus for salt production.

During the latter part of the nineteenth century, a change occurred both in production scale and in technological intrusion. Instead of just gathering salt exuded by the earth, companies started probing more aggressively. In Michigan, the two main methods were mining for rock salt and injection of hot water deep into a local salt layer to produce a brine, which was then pumped back to the surface. By the turn of the twentieth century, both methods had proven their worth so successfully that Michigan was leading the nation in salt production by 1905. It retained this crown—at least for 90 percent of the time—until 1958. Salt mining was even performed underneath the city of Detroit. It must have been worth it, in spite of the difficulties one can imagine.

During the 1890s, the chemical industry also came to settle in the Lower Peninsula. It made use of salt as its main raw material, transforming it into dozens of products: caustic soda and soda (sodium hydroxide and sodium carbonate, to give them their proper names), hydrochloric acid, calcium chloride, magnesium chloride, and so on. The main pioneering companies in those times were the Wyandotte Chemical Company at Wyandotte and the Dow Chemical Company in Midland.

The latter corporation, Dow Chemical, built an empire on the salt supply from the Lower Peninsula. Starting in 1890, it pioneered the use of the chloralkali chain, in which electricity serves to split brine into caustic soda, on the one hand, and hydrochloric acid, on the other (*hands* denote here negative and positive electrodes). In around 1910, Dow began using bipolar electrolysis cells for this process. Sturdy and easy to manufacture, these comprise an asbestos-based membrane enclosed in a vacuum that prevents the chlorine and soda from recombining. This technology, which Dow Chemical has kept proprietary, has a higher output and a lower cost than other competing methods.

Dow Chemical still uses this method today to produce chlorine and soda at the same time; the first of these coproducts helped Dow to build a chloralkali chain that gave it a competitive edge that has made it one of the leading chemical companies, not only in the United States but worldwide. Rival corporations may have their own manufacturing plants also based on brine electrolysis, but their cells are less efficient than Dow's. But allow me to elaborate further on the reasons Dow became one of the leading suppliers of chemicals.

RAW MATERIAL FOR AN INDUSTRY.   Is it commonly known that salt serves as a raw material for building and for the construction industry, that this is even, strictly in terms of tonnage, its primary use, along with its use in the pharmaceutical industry? Are these two claims surprising? PVC (an acronym that, spelled out, means polyvinyl chloride), which is derived both from salt in its chlorine component and from petroleum in its vinyl component, is used abundantly for tiling, roofing, pipes, and flooring, among other things. A good number of medications are made of molecules containing at least one chlorine atom, since that element has the reputation in laboratories of boosting existing biological activity. As early as 1983 nine chlorinated molecules figured in the ranks of the top fifty most-prescribed medications in the United States, Valium and Librium among them.

Chlorinated medications were important players in the Second World War: in addition to chlorophene, an all-purpose disinfectant, and chloroguanide, one of the best-tolerated antimalarial drugs, chloroquine is the antimalarial drug of choice.

Among the tetracyclines, broad-spectrum antibiotics, aureomycin (or 7-chlortetracycline) was the first to be isolated, in 1947. Chloramphenicol, from the same period, cured forty million patients in the twenty years following its introduction.

We owe the recent psychiatric revolution that emptied asylums to chlorinated molecules. Chlorpromazine and other phenothiazines are among the medications most used to treat psychoses; these neuroleptics calm agitated patients. From another chemical family altogether, one that is also chlorinated, haloperidol (or Haldol) is a heavily used tranquilizer.

That salt, used for centuries on end solely for human and animal food, should also have become a raw material for industry is a consequence of basic science, more precisely, of the isolation of the elements sodium and potassium by the young English chemist Humphry Davy in 1807.

Humphry Davy's Discovery. This discovery came at the start of the dazzling career of a young romantic scientist, the darling of all London, a poet when so moved, a genius at popularizing, a pioneer of applied science (we owe the miner's safety lamp to him): Humphry Davy had the insight of turning electricity—that then-mysterious fluid, which his contemporaries took to be a ubiquitous essence, to be found in lightning bolts as in the mass hysterias once orchestrated by Mesmer—into a chemical agent.

This conjecture proved to be inspired, since we now know the role played by electric forces in the structure and cohesion of matter: staying with the example of salt, or sodium chloride, its component atoms, that is, chlorine and sodium, are made of a nucleus with a positive electric charge, an island floating in a sea of electrons, which are negatively charged particles. And if one brings together an atom of each of these elements, the chlorine atom steals an electron from the sodium atom: the first thus becomes a negative chlorine atom, and the second is transformed into a positive sodium atom.

In the solid state, sodium chloride is a crystal, that is, an extremely regular structure—like soldiers on a parade ground lined up as far as the eye can see—of sodium atoms carrying positive charges, alternating with chlorine atoms carrying negative electrical charges. The pattern (sodium +, chlorine −), repeated indefinitely in the three dimensions of space, is constitutive of sodium chloride. Moreover, this is the explanation for its name, since in the lingo of chemists "negative chlorine" is named chloride.

But to return to Humphrey Davy and his discoveries of the elements sodium and potassium: he had at his disposal the brand-new first direct electric current generator in history, the electric battery that Alessandro Volta had invented in Pavia in 1800. When Davy ran an electric current through caustic soda (sodium hydroxide), he was able to isolate sodium from it; he did the same for potassium, isolating it from potash.

Thus one might imagine, given the myriad applications of caustic soda and chlorine (the other element produced in electrolysis, the decomposition of sea salt by electric current), some of which already were well established by the end of the eighteenth century (processing fats with caustic soda to make soap; bleaching fabric using chlorine, which Berthollet had discovered), that Davy's discovery would be followed by rapid industrial development. In fact, it took three quarters of a century for this to happen.

Not that Davy's discovery of electrochemistry was forgotten: on the contrary, between 1820 and 1830, it helped the Swede Jöns-Jakob Berzelius conceive of a dualist theory of chemical matter, in which the electrical attraction between negative and positive charges—the very attraction described above—was posited as the unique law governing the existence of all matter.

Electrochemical Production of Chlorine and Soda. Why did it take such a long time for the electrochemical production of sodium carbonate and chlorine to come about? It wasn't until 1891 that Castner patented an electrolysis cell for the industrial production of sodium from sodium carbonate. There was a comparable lag in the development of a process for obtaining sodium from sodium chloride by electrolysis: beginning in 1833, Michael Faraday had shown the way, but it was only in 1921 that Dupont de Nemours introduced Downs's electrolysis cell in their Niagara Falls plant, where the electric current produced sodium hydroxide from sodium chloride in a first stage before releasing sodium by electrolysis of caustic soda.

One major reason for the delay in the industrial development of brine electrolysis was competition from various methods of chemical reduction. In 1824 Oersted demonstrated that one could derive aluminum by reduction of aluminum chloride using sodium. In 1854 Henri Sainte-Claire Deville and Robert Wilhelm Bunsen developed this into a process, which Sainte-Claire Deville adapted into a method for the industrial production of sodium (used thereafter for thirty or so years) by reducing a mixture of sodium carbonate (produced at the time with the Leblanc method) and lime. Castner improved on this in 1886 by reducing sodium carbonate using iron carbide. But almost immediately the 1888 discovery by Charles Martin Hall of the electrolytic production of aluminum (by electrolysis of alumina dissolved in cryolite) crushed any interest in this way of obtaining aluminum using sodium.

Can one then assert that the electrochemical production of caustic soda and chlorine, since it prevailed over competing methods, after an apparent process of natural selection, led to the best method? At first, it would seem so. In 1893 the Canadian Ernest A. LeSueur opened the first commercial factory for the manufacture of chlorine by means of sodium chloride electrolysis in Rumford, Maine. As noted above, at the start of the twentieth century, the American company Dow Chemical, in Midland, Michigan, began to produce chlorine and caustic soda using brine electrolysis; it is now the industry leader.

Coproduction: A Necessary Evil. It subsequently became clear, however, that the obligatory coproduction of the chemical industry's two chief raw materials, caustic soda and chlorine, was hardly desirable from a strictly economic point of view: in fact, it was rare for demand to match production for both products. Although this did sometimes occur during boom periods of prosperity and expansion, as a rule, the demand for one generally outpaces the demand for the other, which leads to costly stockpiles.[2] Thus both the technological and economic aspects must be taken into consideration. The persistence of the chloralkali chain and of the Dow process is due to rival single-production operations—those producing either chlorine or caustic soda—having remained too costly to the present day.

But salt is not only a raw material for the industrial manufacture of these two commodities—caustic soda and chlorine—produced in very great tonnage (in the hundreds of millions of tons, annually, on the global scale). Other commodities are produced from salt as well.

The Solvay Method.   Consider, for example, sodium carbonate, so-called soda, another major chemical product derived from sodium chloride.

In 1861, in Couillet, near Charleroi, Belgium, Ernest Solvay, then twenty-three years old, began commercial development of the sodium carbonate production method that bears his name. The addition of sodium chloride at first precipitates sodium hydrogen carbonate from a solution of ammonium hydrogen carbonate; the latter is then calcinated to provide sodium carbonate (previously produced using the Leblanc method). Ernest Solvay perfected and developed this procedure for industrial use; it originated with Augustin Fresnel in 1811 (yet another delay in industrial development, in this case, by half a century). The Solvay method is still used to this day, by the multinational corporation of the same name headquartered in Brussels.

In practice, a brine is ammoniated and then carbonated with carbonic gas from lime kilns. These kilns thus also produce lime, which functions to reactivate the ammonia, with secondary production of calcium chloride. This secondary production, of a salt that is actually of little value (calcium chloride), is one of the drawbacks of the Solvay process.[3] Competitive sources of sodium carbonate saw the light during the second half of the twentieth century, with the exploitation of sodium carbonate deposits (particularly in Wyoming), as well as the creation of sodium carbonate from caustic soda rather than from salt.

How is it that salt, this white gold, rare and expensive for centuries, could serve as the raw material for the industrial production of bulk commodities? The answer is predictable: salt's rarity was contrived, faked so as to justify inflated sale prices and taxes. In fact, salt is a very common product, easy to obtain at low cost. It has undergone an extraordinary devaluation since the nineteenth century, which saw the start of its industrial use and development: this essential product is not worth much anymore, only somewhere around $0.40 per kilo. Should one thus conclude that chlorine and caustic soda are also inexpensive products? Absolutely not: in spite of their enormous production volumes, the prices of such raw materials for industrial purposes can skyrocket every time the market loses elasticity.

Consider the example of caustic soda: in the third trimester of 1997, its price soared, reaching three successive highs in three months. The price rose from $90 per ton to settle at $170 per ton, the caustic soda being supplied by its principal producers: Occidental, Olin, and Elf Atochem, along with Dow Chemical.

The reason for this impressive increase was strong demand, spurred by the economic recovery in the United States and other industrialized nations (Great Britain, the Netherlands), coupled with a supply at maximum capacity, with the result that the slightest incident would send prices soaring. In this instance, the unexpected stoppage of Dow Chemical's chloralkali facility in Fort Saskatchewan for an entire six weeks (its production capacity is 720,000 tons per year) sparked this spectacular increase.

It must be said that chloralkali producers, facing bans and reductions affecting chlorinated products (I will get to this later), are hardly investing any longer in new production units or even in increasing production capacity. The paper industry was a big outlet for them, but, following the revelation of traces of hypertoxic dioxins in paper factory wastewater, chlorine is well on the way to being banned from this industry, where it had been used to bleach paper pulp (other methods have supplanted it, in particular those using ozone or oxygen). At this point, a closer look at the innocuity of chlorine-containing chemicals is merited.

The Janus Face of Chlorine. So, what about these toxic molecules? If they contain chlorine, don't they become harmful and dangerous? Among major polymers, the monomer used in the manufacture of PVC, vinyl chloride, is a powerful carcinogen; great precautions must be taken by workers to avoid inhaling them. Chlorinated organic solvents, such as chloroform, carbon tetrachloride, methylene chloride, and trichlorethylene, that have enjoyed wide use—not only in the chemical industry but in dry cleaning, for example—are themselves also carcinogens, toxic to the liver and the spinal cord. Chlorofluorocarbons or CFCs (Freon) served as liquid refrigerants and propellants for aerosols. Over the past thirty years, it has been confirmed, thanks to the work of the Nobel Prize–winning chemists Sherwood Rowland and Mario Molina, among others, that chlorofluorocarbons were to blame for the hole in the ozone layer: In the stratosphere, under the effect of ultraviolet solar radiation, chlorine atoms are detached from their molecules. Each chlorine atom thus freed destroys as many as ten thousand molecules of ozone in the thin shield that surrounds our planet.

Other types of chlorinated chemicals are more harmful, even much more pernicious. This is true of two large families: polychlorinated biphenyls, or PCBs (long used, among other things, as electrical insulation in transformers), and dioxins. The latter, even completely banned from the chemical industry, nevertheless continue to be found in trace amounts in the natural environment. Among the many processes that give rise to dioxins are cases when chlorinated derivatives come in contact with organic matter. Examples of this are, for chlorine, paper pulp bleaching (now performed using other methods); municipal incinerators burning household refuse that contains both chlorinated products, such as salt, and organic matter (this also creates airborne ashes—fly ash—that catalyze the dioxin-producing process); and chlorination of municipal water supplies to destroy germs (typhoid has been eradicated in this way since the start of the twentieth century) when groundwater has become contaminated by organic substances. PCBs and dioxins in even minuscule amounts are violently toxic.

Such negative aspects explained strong action against chlorine and chlorinated products on the part of ecologists, such as Greenpeace militants, who sought to ban entirely their industrial production and household use. Dow Chemical's chlorine chain, for example, has been the object of attacks these last twenty years by various ecological movements. This company has carried out a skillful response, improving its public image by lowering noxious wastes considerably, cleaning up the areas surrounding its factories and taking advantage of this pressure to diversify and find new markets, all without suddenly renouncing its chlorine chain.

In any case, things are not so simple. Chlorine and chlorinated compounds yield as many beneficial products (for drinking water treatment, for instance) as harmful ones (dioxins and PCBs). Nor can one allege either that the artificial seeks out and destroys the natural, for, to take but one example, there are hundreds of chlorinated natural substances in ocean organisms. And human ingenuity can rival nature for sheer diversity of productions: consider, for instance, all the uses vinyl chloride was turned to.

THE AGE OF VINYL.   Salt is the primary material for the manufacture of chlorine. And chlorine in turn has for its main application—after its combination with ethylene, a product from oil refineries—the making of vinyl chloride. Finally, vinyl chloride is turned into the polymer polyvinyl chloride, PVC for short or "vinyl" as it is often termed.

Vinyl will always remain emblematic of the United States of America during the 1950s, at the time of the Eisenhower presidency. This was a country that, during those blissful years, had found for itself a fragile equilibrium of poise and counterpoise, between idealism and smug self-satisfaction, between pragmatism and intellectualism, between the materialistic pursuit of comfort in its numerous attractions and the self-assertion of the military victors of World War II.

The Vinyl Era was epitomized when Nixon went to open the American National Exhibition in Moscow and argued publicly with Khrushchev, on July 25, 1959, for the superiority of the American way of life in their famous "Kitchen Debate." Vinyl had become a fixture of American life: it was the material of the everyday, the ubiquitous stuff that spelled out in hundreds of embodiments "Easy does it."

The kids would rampage through the kitchen and drop a jar of peanut butter on the floor. Mom would later come along and wipe clean the vinyl tiling. Teenagers would abscond at night to a drive-in theater. Their embrace on the back seat of the car would press their sweaty bodies against the vinyl upholstery, mixing up odors of plastic and antioxidants with more animal smells. At home, they would congregate to the sounds of big-band jazz or Latin American dances such as the samba recorded on long-playing 33 1/3 rpm vinyl records.

Clothing styles were changing. Shoes, for instance, for both men and women, witnessed the end of the exclusive domination of leather and styles emulated from the British. Much less expensive sandals and pumps, which Italian designers were crafting from vinyl with imagination and flair, were the easygoing footwear that seemed natural to young adults besotted by the fad of the vinyl hula-hoop.

The attendant growth of metropolitan suburbia, if one surveyed it from an airplane, was seeded by vinyl. One could see the near-epidemic pasting of tiny blue disks on green backgrounds, reminiscent of colonies of microorganisms spreading in a petri dish, as swimming pools in the backyard became the norm. Lawns were kept fresh with sprinklers, and vinyl tubing supplied water to meadow and pool alike.

The Howard Johnson motel-restaurant combination became in the Age of Vinyl one of the most familiar sights on the American landscape. Each advertised on a giant billboard two or three dozen available ice cream flavors. Howard Johnson's thus became emblematic of a new incarnation of the neighborhood diner, with a stereotyped look and mass-produced menus and food, with identical vinyl-covered stools and booths stretched across the length and breadth of the continent, with the same orange vinyl-tiled roofs, and with uniform parking lots populated by Fords and Chevrolets, Studebakers and Buicks, De Sotos and Cadillacs, and Chryslers often sporting the same colors as the scoops of ice cream being served inside, in vinyl containers and on vinyl mats.

The lesson from vinyl seemed to be: You can enjoy uniformity with apparent diversity; everyone is the same although different, and prosperity can be had by all; there is "better living through chemistry" (the then-motto for DuPont de Nemours); and vinyl is the wonder material, durable and yet disposable, clean and inexpensive, abundant and rich-looking, obsolescent and renewable, that allows the American way of life to thrive, endure, and propagate. One of the features of this way of life most remarkable to a foreigner such as myself entering the United States in the early sixties was how pleasant and widespread air conditioning was. Interestingly, it too can boast that salt played a role (and what a role!) in its history.

SALT AND COLD. One summer day in 1620, James I and his courtiers in Westminster Palace were very hot. The king had chosen this day to attend the demonstration by the Dutch inventor Cornelius Drebbel of a number of machines he had built. One of Drebbel's inventions, an air cooler, so drastically lowered the temperature in the great hall that the king and his retinue fled shivering. Drebbel had used salt to lower the temperature of the water.

One finds a trace of this public demonstration in the writings of Drebbel's illustrious contemporary, Francis Bacon. In *Novum Organum*, published that same year, he wrote: "Nitre (or rather its spirit) [we translate this as: potassium nitrate or saltpeter (or nitric acid)] is very cold, thus nitre or salt added to snow or ice heightens their coldness [. . .], salt does so by adding to the activity of the snow's cold."[4] And in his own Latin translation of *The Advancement of Learning*, written in 1622 and published the following year, Bacon alluded to Drebbel's public demonstration,[5] while in the posthumous *Sylva Sylvarum* he again returned to the artificial production of ice with the use of salt.[6]

How can the presence of a salt, or of salt, lower water temperature sufficiently to produce ice? First, recall that the melting of ice to make liquid water is called a "change of state," as is the vaporization of that liquid to form a gas, namely, water vapor, and that the temperatures for these changes of state for pure water are 0 and 100°C. It is not surprising that these two temperatures would be altered by the introduction of an impurity, such as salt dissolved in the water.

If one starts from the principle that the melting and vaporization temperatures of salt water are not the same as those of pure water, can one predict whether they increase or decrease?[7] To find an answer to this question, let me outline an argument. I base it on the following insight: the molecules that make up a gas are more disordered than those that make up a liquid, and those in a liquid are more disordered than those in a solid; a snapshot of their relative configurations would reveal no change in a solid, rapid variation in a liquid, and even faster variation in a gas. One can thus assert that each of the changes in state—solid to liquid and liquid to gas—increases disorder, which can be observed on the molecular level.

In order to clarify further this notion of molecular disorder, assume that one could examine, with the help of a microscope, the composition of a sample of salt water at any level whatsoever, moving from the scale of a glass of water to that of an element of microscopic volume. In carrying out such a "zoom" viewing, one would first see a homogeneous liquid medium with what appeared to be the same properties at every point. However, with magnification from the microscope, one would discover that the liquid sample under the lens, depending on the moment it is examined and also on the sector of space being explored, contains or does not contain dissolved particles of sodium chloride. Homogeneous on a large scale, the liquid is no longer so on a small scale because of the presence of dissolved salt. Figure 3 reproduces a diagram of this: each of the boxes (small cubes, actually) represents a volume unit of liquid, pure water on top, salt water below. All the boxes are the same for a pure liquid, whereas they are no longer necessarily all identical in the case of the mixture: some boxes, marked with a cross, now contain salt.

Imagine that there are only nine boxes to consider, as in figure 3 (in reality, there would be billions) and that these must either remain empty of salt (pure liquid) or hold, for instance, two salt particles (salt water): the first situation, shown at the top of the figure, is monotonous: it always remains identical with itself. The second situation permits no less than seventy-two different possible outcomes: once the first particle of salt is introduced in one of the nine boxes, there remain eight more boxes in which to place the second particle (and nine times eight equals seventy-two).

In short, this thought experiment reveals great anarchy on the microscopic level, since all sorts of volume configurations coexist: viewed in this way, salt water is more disordered than pure water.

But my question had to do with the change-of-state temperature, and in changes in this temperature depending on whether the water is pure or salty. Consider the melting of a solid: most certainly, as I have said, the resultant liquid is more disordered than the solid, since its molecules have gained a freedom of motion that was denied them in their

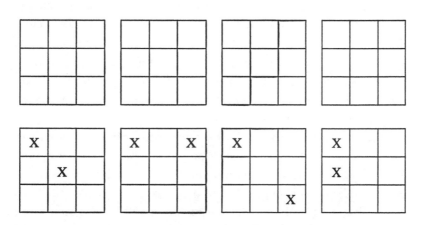

FIGURE 3. Diagram of a liquid, pure on the top row, mixed on the bottom row. In the top row, cells in a microscopic sample are all identical. In the bottom row, a number of configurations coexist (the X indicates an impurity).

solid state. Moreover, a certain amount of energy had to be supplied to the solid in order to disorder it and to be able to observe its melting— by heating it (or, in what amounts to the same thing, by borrowing thermal energy from the outside environment). Since salt water is more disordered than pure water, less thermal energy will be required to melt the solid. Thus the melting point for ice will be lower when salt is present. Conversely, the boiling point of salt water will be higher than that of pure water, but this second point is of no further concern here.

This little argument traces a law of nature that the Grenoble-based scientist François Raoult was able to confirm and quantify at the end of the nineteenth century: any impurity in a solution lowers the melting point of the attendant mixture and raises its boiling point, and the variation in the change-of-state temperature is proportional to the number of microscopic particles of the impurity present. Less energy (from heating) thus needs to be supplied in order to melt ice into salt water, which is more disordered. In the opposite direction, since the salty aqueous solution is already more disordered, it is less inclined to vaporize, and more calories need to be supplied for boiling to occur.

In effect, all that is needed to create a refrigerant mixture, that is, an aqueous system whose melting point is lowered from 0°C to temperatures (depending on the salinity of the liquid) between about −15 and −20°C, is to add ice to abundantly salted water. You can easily verify this fact with a thermometer, and this little trick will also permit you to chill your champagne bucket more quickly! There are numerous other such applications; for example, ice skating rinks are salted so that less energy is required to transform the water into ice. Snowy roads are also salted, taking advantage of the same phenomenon but for different effect. As I have just shown, in the presence of salt, water and ice coexist at temperatures well below 0°C. Once a road has been salted, then, water stays liquid and thus forms no ice or black ice even at very low temperatures. This remains one of the main commercial uses of salt.

Salt and cold may make common cause at times, but they are rivals when it comes to the preservation of perishable foods. Salting permits, and has since time immemorial, the preservation of proteinaceous foods such as cheese, fish, and meat; likewise, refrigeration also prevents bacterial proliferation. Toward the middle of the nineteenth century, with the advent of facilities for the industrial production of ice (through use of the steam engine), refrigeration gradually took over from salt the role of fish preservation on fishing boats and on the

boats and vehicles transporting fish to urban centers of consumption.[8]
Since the time of the French Revolution, Scottish fishermen had
refrigerated the salmon they delivered to London's Billingsgate fish
market using salt. If salty water can be cooled to lower temperatures,
other, seemingly mundane phenomena also beg to be explained.

SALT AND WATER.   My aim here will be to show how scientists turn
an apparently simple reality into a problem for investigation. Consider
the following three observations: if I put a bit of salt in water, it dissolves ("it melts," as children say). Conversely, if I evaporate the water
from a salt solution, the salt crystallizes. Finally, the ocean's salinity, not
very far from the maximal salinity of a brine, is thirty-five grams per
liter. Are these facts?

The scientific approach consists in being astonished, even amazed.
First questions: How does the dissolution occur? How can it begin to
take place? The answer is not obvious, because the crystalline edifice
of salt, constructed of sodium and chlorine ions, is a sound one. For
it to split apart in water, the water molecules have to bestow on the
sodium ions, which have a positive electric charge, and the chlorine
ions, which have a negative electric charge, water-sodium and water-
chlorine forces of attraction equal in strength at least to the mutual
sodium-chlorine attraction. This is plainly the case: one can show by
scientific calculation that these water-sodium and water-chlorine
forces are indeed strong enough.

The second question, concerning the initial phase of dissolution,
can be posed in this way: how is it that a pinch of salt in a small spoon,
exposed to the surrounding humidity, does not spontaneously and
instantaneously dissolve into a salt tear? In other words, salt crystals
seem invulnerable to water molecules in the atmosphere, though they
dissolve immediately in liquid water. We can say that this system composed of salt crystals and water vapor is metastable, that is, it persists
when this same system would be more stable in the form of a mixture (salt water).

To explain this slowness, this difficulty in dissolving, imagine the
first water molecule that comes along, by diffusion in the air, and collides with the crystalline solid NaCl. Suppose that it turns its positively
charged part toward the salt; instantaneously, an attraction between
this water molecule and the negatively charged bits of the salt, namely,

the chlorine ions, establishes itself. The nearest chlorine ion would then attach this first water molecule to itself. But it is flanked by the positively charged bits of the salt, namely, the sodium ions. The water-salt conjunction is thus frustrated by repulsive electric forces that oppose the attractive electric force. If the water molecule had attacked the salt crystal from the other end, the reverse would have occurred: a water-sodium attraction but accompanied necessarily by repulsive forces between the water molecule and the chlorine ions.

The foregoing thought experiment explains the presence of an energy barrier separating the two states of the system, the initial metastable state and the final stable state. In practice as in theory, time, as well as innumerable instantaneous water-salt collisions, would be necessary before the water molecules could manage to fissure and shatter the sodium chloride crystal.

Whether one investigates it in thought or views it on a microscopic scale with the help of instruments, little by little this dissolution takes on the appearance of a tiny catastrophe. In fact, the attractive electric force $(+, -)$ is diminished a priori by a factor of about eighty in liquid water compared to that in air because of the differing electric (dielectric, to be precise) properties of these mediums; this is one of the reasons for the apparently instantaneous dissolution of salt in liquid, while dissolution is very slow in humid air. Thus, as soon as minuscule pools of water have begun to form at the heart of a salt crystal, it will weaken, and its dissolution will only be accelerated by this.

Now for the converse problem, namely, that of crystallization. How does it begin? With the sodium ions $(+)$ and the chlorine ions $(-)$ in the salt solution coming and going in all directions, each in a protective shell of water molecules, crystallization entails an entire series of events:

- loss by a hydrated sodium $(+)$ ion of at least one water molecule from its protective shell;
- loss by a hydrated chlorine $(-)$ ion of at least one water molecule from its protective shell;
- the collision of these two entities, the interpenetration of the water layers, and the gradual joining of the partially dehydrated sodium and chlorine;
- and it starts over again, in such a way as to increase the size of this crystalline seed.

In other words, the problem is that of complementary particles recognizing each other on the microscopic level. This issue of recognition on the microscopic level is one of the current areas of major interest in chemical science, very much influenced in this respect by contemporary biology.

Why can one not dissolve more than a certain amount of salt in water? When a solution becomes so saturated (as it is called), one witnesses the spontaneous crystallization of the excess amount of salt. A brief calculation shows that at the oceanic salinity level of thirty-five grams of salt per liter of water, the average distance in the solution between the individual chlorine and sodium ions is on the order of three or four times the size of these corpuscles, swollen as they are by water molecules clinging to them in their protective shells.

How can sodium/chlorine recognition—the prerequisite for their conjoining and thus for the gradual increase in the attractive force that acts to unite them—take place? A first answer lies in their distance from each other: if this distance is an average of fifteen (no matter the unit of measure), certain sodium-chlorine pairs will be located at lesser distances—of ten or even eight—and thus these pairs will give rise to the first conjunctions. In this case, there is (once again) a statistical aspect that links corpuscular interactions on the microscopic level to the macroscopic event of crystallization.

A second answer, which incidentally does not rule out the first, stems from the transmission of electrical forces through the intermediate water molecules. This polarization, a localized factor, counterbalances to a certain degree the aforementioned reduction by a factor of eighty of the electrostatic forces in effect in liquid water, a reduction that is a generalized phenomenon. But let me turn now from science to superstition or, rather, to a mix of science and superstition.

THE WINE STAIN.   A glass of red wine is tipped over, jostled by a careless gesture. A large stain begins to stretch out across the tablecloth. But a guest quickly intervenes, sprinkling salt on the fabric. The liquid, which has already spread, is gathered up into a thick, whitish, pasty spot, a little bit pink, that a knife blade can now lift off of the material.

Replace this clichéd scene—how else to characterize such a familiar, everyday sequence?—with scientific analysis. Scientific analysis is astonished; it poses questions at the point when common sense is content to recognize something as fact, to assimilate it as normal. What would have taken place in this case if salt had not been used? The liquid overturned on the tablecloth would have stained it, because the liquid portion of the wine would have passed through the fabric's weft, depositing the natural dyes that account for its color on the fibers in the process.

Why is this? It is because strong attractive forces on the microscopic level are in effect between the fibers in the tablecloth—fibers designed for cloth made of cotton cellulose (a glucose polymer, made up of innumerable units, each made of a molecule of this sugar, strung together in a single chain)—and the dye molecules.

In other words, the fact that the dye molecules fixed onto the cotton indicates their attraction for the cellulosic fibers, an attraction at least as strong as the attraction they showed for the water the moment before, when they were dissolved in it as wine. It's as though the cellulose fibers were fitted with tiny Velcro strips that hook and capture the dye molecules.

The whole danger of the "cheap red wine that stains" is to be found here, in the strong fixation of molecules with each other. Once the tablecloth is splattered, it has to be laundered in plenty of running water and even with a modern detergent—formulated precisely to combat the consequences of this type of scenario—to rid it of the stain.

Salt in this case acts as a helpful aid. But before I examine the interactions that take place on the molecular level in order to explain the physicochemical phenomenon involved, one point must be mentioned.

Actually, those fictitious strips of Velcro on the surface of the cellulose are capable of fixing a good many other molecules in addition to the natural dye molecules in wine, just as an art gallery can hang paintings of the most varied sizes and styles. These microscopic adhesions that intervene between the support base and a diverse group of molecules bear the generic name of *adsorption*. Let me mention two applications of adsorption before returning to the use of a pinch of salt to prevent the staining of the tablecloth.

Paper collages are among the works produced by an artist of my acquaintance. She makes her own paper pulp from powdered cellulose derived from cotton; all she needs to do is add a little water to it. She then introduces various natural or synthetic dyes into this aqueous, rather viscous medium made of a suspension in water of cellulose corpuscles joined to each other in a three-dimensional network (or gel). The dye molecules migrate to the cellulose fibers and lodge there permanently: they are adsorbed there, so to speak. This lasting adsorption is what colors the resulting paper once the water has evaporated. You now know how one obtains sheets of paper—or in the case of this artist, paper forms—that are colored.

The second application I want to discuss is a laboratory experiment assigned to a good many beginning students. After crushing spinach leaves in a mortar, they extract the green colors from the crushed matter with a solvent such as acetone or chloroform. And the color in the spinach leaves comes not from chemical substances that are all identical but from a mixture of several substances; among these are carotenes (the orangy yellow or red molecules that lend their color to carrots and tomatoes) and chlorophylls (which are responsible for green plants' conversion of solar radiation into atmospheric oxygen and molecules of the living organism).

But how can the students carrying out this experiment determine with certainty that they are dealing with a mixture of dyes yielded by the spinach leaves? They place the same cellulose powder used by my artist acquaintance into a glass tube plugged at one end with a liquid-permeable stopper. They then pour the colored solution, into which part of the green from the leaves has passed, into the tube.

What happens then? As one might predict, the dye molecules are adsorbed onto the cellulose molecules: the little Velcro-like microstrips have once again done their duty. The experimenters then pour various liquids on the column of cellulose, held upright; among these liquids are solvents selected to produce interactions with the dyes that are at least as strong as the cellulose-dye attraction. These influxes of liquids unhook the dye molecules from their temporary cellulose support base, but only after the adsorptive forces, which are a function of the particular chemical structures of the various dyes, have carried out their sorting process. After leaving the column, the population of molecules explodes into subgroups—which can be seen with the naked eye—comprised of the various carotenes and chlorophylls present. This experiment illustrates the fundamental principle of chromatography, the most widely used separation method in laboratories today.

All that remains is to explain in a few words how salt prevents adsorption of a wine spill's colors by the tablecloth. Salt dissolves in water and thus in wine, an aqueous mixture. The small crystals of the sprinkled salt attract the nearby wine, with its dissolved dyes, in such a way as to dissolve in part there as well. The other salt crystals, those that remain in a solid state, give the mixture its pasty appearance and adsorb some of the dyes on their surfaces as well. Actually, it does not work too well: only a minute fraction of the wine is thus prevented from staining the tablecloth. The custom is more superstition than empirical success.

SLIPPAGES.   Work in my laboratory is done with the help of clays: after we have adapted them for a given purpose, they serve as catalysts for the reactions of organic chemistry. From countryside to test tube, from stone to mineral, from the scale of the kilometer to that of one-millionth of a millimeter, clays retain certain features, certain behaviors. Isn't this remarkable?

One of the most familiar properties of clay is its plasticity, the feature a potter uses to center a ball of earth and throw a pot. Once again, the explanation lies on the microscopic level, on the scale of one-millionth of a millimeter: all clay is made of platelets of an aluminosilicate. These platelets can pile on top of each other, like a deck of cards. Ordinarily, when not dried out under the sun or in the studio, these platelets are separated by layers of water. Thus two adjacent platelets in a stack have one or more layers of water molecules interposed between them, most often layers of salt water in natural clays (there exist artificial clays as well). This lamellar structure is what gives clay its plasticity: each of the platelets—like a card in a card deck, or like a record, or even like a book on a shelf—slides easily in relation to the others under the effect even of weak forces. The layers of salt water between the platelets form a kind of carpet, on which any lateral movement is easily accomplished.

Such slippage affects every sample of clay. This is equally true for landslides in clayey landscapes (even huge hydroelectric dams have suffered such catastrophes), for the model for a future terra-cotta piece in a sculptor's studio, and for the vase thrown on a potter's wheel, as long as there is water there—salt water, remember, as a general rule— to serve as a lubricant. Clay that has been completely dehydrated loses its plasticity and becomes a powder or, at high temperatures, a ceramic. But, unless it has been fired in a kiln, one need only moisten it with water for its sheets to recover the ability to slide microscopically over each other and, the equivalent of this on the macroscopic scale, the ability to be molded.

The salt that is present helps to preserve the humidity at the heart of the clay. Salt occurs there broken down into sodium and chlorine ions, each one of which strongly attracts water molecules to itself. Moreover, salt—that is, sodium chloride—can be replaced by other salts, by calcium chloride in particular. What might this be good for?

WATER SOFTENING.   Hard water—which deposits scales of calcium carbonate on the bottom of pans and kettles, makes it difficult for soap to lather properly, and so on—is defective. To soften it, that is, to rid it in large part of its calcium, all one need do is pass drinkable water through an ion exchanger. Clays can function as such ion exchangers, and in fact were the first ion exchangers used to soften municipal drinking water.

How does this softening take place? When water passes through clay, the sodium ions in the clay are exchanged with the calcium ions in the water being treated; the water leaves the clay softened by the sodium ions that have replaced its original calcium ions, while the clay is in turn enriched with calcium. To reactivate the ion exchanger, one has only to add sodium to it in the form of ordinary table salt (or sodium chloride), which expels the calcium. After that, it can be used again or discarded in the environment (it is harmless). Those who use dishwashers are familiar with such calcium-sodium exchangers, which are continuously self-renewing and allow glassware to remain clear.

But one must not conclude from this that the types of ions that may lie between the sheets of a clay are limited solely to sodium and calcium. A good many other positively charged metallic ions—potassium, for instance—can be made to enter and lodge in the clay.

An example that is both amusing and useful is that of organic cations, known by the archaic name of quaternary ammoniums (they are also used in detergents, for shampoos, among other things), which are symbolized with the notation Q^+. These Q^+ cations are easy to manufacture and very inexpensive, and they act—since their structures include chains of hydrocarbons—like so many particles of oil inserted into the aqueous environment between the clay sheets. When one replaces the sodium at the heart of a clay with Q^+ cations, one obtains a compound, a hybrid substance joining the clayey mineral with the Q^+ organic cation. These compounds, called organophilic clays, have numerous uses in the food industry, among others. Organophilic clays, for example, lend ketchup its thyxotropic quality: viscous at rest, it liquidizes with a shake. The explanation for this singular physical behavior is that the clay sheets slide over each other even more easily when Q^+ cations are present to provide lubrication between the clay sheets. This makes geophagists, that is, earth eaters, of us all. But true geophagy is another story, one about potassium rather than sodium, and so it therefore does not belong here. What does belong here is a reference to the prowess of art in making us forget our mortal condition and admire its mastery.

SALT GLAZING.   I opened this chapter with an example of the union, found in seventeenth-century alchemy, of traditional thought and rational, prescientific thought. I revisit this theme now in making reference to a technique used in stoneware ceramics that has its origin in Zen Buddhism.

To this day, a Kyoto stoneware potter, at the end of a firing, will throw a handful of coarse salt over his pieces. This gesture, somewhere between resolute and nonchalant, stamps the work with the seal of supreme artistry, an artistry of chance, of the unexpected, and of the serendipitous.

Salt glazing is also done, especially in the Germanic and Anglo-Saxon countries, by soaking a piece in salt water before firing it or by coating it with a sodium compound such as crystals of caustic soda or sodium bicarbonate.

Originating in the Far East (Korea, Japan), there is evidence of stoneware salt glazing in the West (Cologne) since the twelfth century, and seventeenth-century technical treatises mention it. Its effect is to cover the stoneware so treated with a very fine, bluish coating (0.1 millimeter thick) that is shiny but flecked, "flambé," as potters say. This thin layer of sodium aluminosilicate results from a chemical reaction between the salt vaporized by the oven temperature (1,300° C), the clay, and the ambient water vapor.

The process remains exceedingly delicate. A technical manual specifies, among other things, that "an insufficient amount of salt yields a mediocre glaze, while an excessive amount, or a temperature that is too low, may cause a white foam to form on the surface of the piece;

cooling the piece too quickly can have the same effect. . . . If the piece is too porous, salt vapors penetrate into the fired piece and produce a dull glaze. On the other hand, complete vitrification of a piece creates streaks. . . . At the point when salt is introduced . . . the fires must blaze intensely."[9]

The precision of the Japanese potter's glance and gesture remains without equal. The accuracy of spectroscopy in identifying the nature and the origin of almost any specimen is also peerless.

THE INVENTION OF SPECTROSCOPY. Robert Wilhelm Eberhard Bunsen and Gustav Robert Kirchhoff, University of Heidelberg professors of chemistry and physics, respectively, were strolling along the Neckar River toward the close of the 1850s. According to one version of the anecdote, Kirchhoff is said to have remarked that incandescent vapors were capable of absorbing the very radiation they emitted. Bunsen replied that he was crazy ("Gustav, du bist verrückt"). On returning from their walk, Kirchhoff constructed a rudimentary device that proved he had been right. Very excited, he went to see Bunsen, exclaiming: "Ich bin verrückt, ich bin verrückt."[10]

According to another version of the story, recounted in the January 12, 1889, issue of the *New Zealand Herald*, on one of their walks Bunsen is supposed to have said: "Kirchhoff, we must discover something so simple that it cannot be true." Their subsequent work in fact led to the aforementioned law regarding the absorption and emission of radiation by matter raised to a high temperature.

We now know that this is indeed a fundamental law, and it allowed Bunsen and Kirchhoff to invent spectroscopy, that is, a way to characterize atoms and molecules based on their absorption and emission of radiation. In all likelihood, however, its first discovery was not made by Bunsen and Kirchhoff.

As we now know, Max Planck's quantum theory explains why the absorption and emission of radiation have identical wavelengths, which was the law supposedly discovered by Kirchhoff. When an atom is stimulated by an elevated temperature, it can lose energy by emitting a photon, or a unit of light. But the energy states of an atom are like the steps in a staircase, and the energy radiated with the emitted photon very precisely measures the distance between, say, in this case, the third and second steps. Thus that same atom at the energy state of the second step can be restored to the energy state of the third step by absorbing a photon of the same wavelength as the one that knocked it down a step in the first place.

But what about the applications of Kirchhoff's law? First, let me go back in time to a few years before the historic walk of the two friends and colleagues along the banks of the nonchalant little river.

Bunsen, a professor of chemistry at the University of Heidelberg (one of the most prestigious German chairs in this discipline) since 1852, was an excellent researcher and what one could call an inspired tinkerer. In 1855 he designed and built an improvement on Faraday's gas burner by adding a rotatable, perforated ring valve, making possible the control and regulation of the entrance of air. This is what has ever since been called the Bunsen burner. With its inexpensive, sturdy flame, it provided laboratories with a heating device that allowed for easy control of the height and intensity of a flame at the flick of a finger.

Having perfected this burner, Bunsen made another discovery, this one methodological. When Bunsen cast some salt, that is, sodium chloride (or any other sodium derivative), into the burner's flame, the flame became an intense yellow color.

Newton had demonstrated the power of a prism to disperse any kind of light: this simple optical instrument breaks any light into its component parts; it allows the component wavelengths to be displayed and exposed to sight. When Bunsen examined the sodium's yellow light using a prism, he noticed that the light revealed only a single wavelength; in other words, it is nearly monochromatic (to use today's terminology) (in fact, it is composed of two very close wavelengths, termed sodium D lines). From then on, this sodium D radiation provided a simple and inexpensive source of monochromatic light, with a lovely yellow color, and did so for a solid century, until the invention of the laser.

This yellow light—you're familiar with it, since some intersections and highways, particularly in Belgium, are lit at night by sodium vapor lamps—is characteristic of sodium: if one throws a sample of an unknown substance into a flame and then sees this yellow light, one can be certain that this sample contained sodium in one form or another. This type of analysis is called spectrum analysis or spectroscopy.

The spectroscopist's mission resembles that distinctive talent of musicians who, based on timbre, are able to detect by ear the various instruments in an orchestra: clarinet, bassoon, double bass. As the conductor easily distinguishes between viola and violin, so the spectroscopist likewise can tell the difference between methanol and water in a mixture, whether that mixture is the cosmic interstellar medium, a biological cell, or an industrial composite.

In order to make such distinctions, the spectroscopist makes use of spectra, as Monsieur de la Palisse [Monsieur de la Palisse was a sixteenth-century soldier who became the subject of a popular humorous ditty characterized by comic truisms. This is the source of a French word for *truism: lapalissade.*—Trans.] might have put it. Each of these spectra, formed by the emission or absorption of radiation at certain characteristic frequencies, is like a bar code for the chemical substances responsible for that emission or absorption; the term "spectral signature" is even used.

Back to Bunsen. His friend Kirchhoff, who had been his colleague at Breslau (Wroclaw today), had himself been named a physics professor at Heidelberg in 1854. The two friends collaborated, using the Bunsen burner to produce emission and absorption spectra for which Kirchhoff had just determined the laws, according to the anecdote I have reported. They thus invented spectroscopy and beginning in 1860 applied it to discover the element cesium (from 40 tons of mineral water) and the element rubidium the following year (from 150 kilograms of the mineral lepidolite).

With his gas burner, Bunsen had perfected a technology; this technology allowed him, in collaboration with Kirchhoff, to advance science through such discoveries and above all through the establishment of spectrum analysis, which enables one to determine the composition of a mixture and the substances found in it.

One of the great advances in subsequent spectroscopy has nothing to do with sodium chloride. A self-taught British astronomer, Joseph Norman Lockyer, trained himself as a solar spectroscopist. In 1868 he observed in the spectrum of solar prominences the spectral lines (bar codes) of an element then unknown on Earth that he named "helium," after the Greek *helios,* Sun.[11] Like many other great scientific discoveries, the discovery of helium was multiple: the Frenchman Pierre Janssen, the founder of the Meudon Observatory, made it simultaneously during the 1868 solar eclipse. Later, terrestrial helium reserves were discovered.

But to return to sodium and to another advance in its spectroscopy, following Sputnik, some French astrophysicists, under the direction of Jacques Blamont and at the instigation of André Danjon, carried out research projects investigating upper atmospheric currents by launching probe rockets. These rockets ejected sodium vapor. The sodium atoms then absorbed the solar radiation, to reemit part of it in fluorescence, with that characteristic beautiful orange color.

Just as a jet aircraft is followed by a water vapor trail that is generated by kerosene combustion in the engines and condensed by the low temperatures of the atmosphere at the high altitude at which it travels, so the sodium ejected by Blamont's probe rockets left traces that gradually lost their shape under the influence of nearby winds and "somewhat resembled the form of a small intestine, with the rounded zigzags reflecting changes in wind direction."[12] This kind of technique made it possible to measure the force of these powerful winds and to map them.

As the converse to Lockyer's discovery of helium, starting in 1930 spectroscopists discovered the presence of molecules in outer space through the work of pioneers such as Gerhard Herzberg (Nobel Prize in chemistry for 1971) and Pol Swings (the father of astrophysics at the University of Liège). Today, nearly one hundred molecules have been identified that are present throughout space, such as water, formaldehyde, and hydrocyanic acid. This has been accomplished chiefly by means of radioastronomy, that is, with spectroscopy based on radio waves emitted by molecules while spinning and detected by radiotelescopes.[13] This discovery is of tremendous importance in intellectual history; it is akin to the Copernican revolution. From that point forward, we know that to restrict life's origin to Earth alone is very probably a geocentric illusion, and that at least in theory we must decentralize the biotope from Earth to the universe as a whole (but only to include the rare places where life forms can thrive).

Two points remain to be made. Like most discoveries, Bunsen and Kirchhoff's discovery is indebted to a good deal of research that preceded their own. In fact, the English physicist George Gabriel Stokes, who held the Lucasian Chair of Mathematics at Cambridge beginning in 1849 (Newton's chair, now held by Stephen Hawking), had himself also discovered spectrum analysis. His correspondence— from before 1852—with William Thomson, another English physicist, provides evidence of this. It actually wasn't until 1859 that Kirchhoff published a report of his discovery. The priority of Stokes's discovery is clear, but he hadn't published. "Publish or perish," as the saying goes.

The second point concerns the anecdotal aspects of the history I have just recounted, for they exasperate serious historians of science. Those professionals have little use for legendary tales of the kind internal to a discipline and absent from written records. In my view, however, they are part of an oral history that is not entirely devoid of relevance and that, on the contrary, should be recorded before it dies out; it also at least has the advantage of making it easier to popularize great discoveries.

To conclude, let me suggest a point of reflection on scientific research, namely, that the revealing of truth rarely treads the path of discovery. It most often follows a faltering and choppy course, which can be frustrating. I will illustrate this by means of a comparison.

VARIATION ON THE SAME OLD TUNE. We're all familiar with this kind of word game: "Come out and play . . . play ball! . . . Baltimore . . . morning star . . . star of David . . . video game . . . game board . . . bored to tears . . ." and so on. Many scientific explanations bring this kind of word game to mind: Why do I see? Because photons strike the retina. Why does the retina absorb photons? Because it contains a photon-trapping molecule, retinal. How does retinal trap photons? By changing shape each time it has ingested one. Why does retinal's change of shape communicate this fact to the optic nerve? Because, in the process, chemical reactions release a proton when retinal attaches itself to opsin, a protein in visual purple. And how does the freeing (or the fixation) of a proton induce a signal in a neuron? Because it orders sodium channels, located along the length of a neuron's axon, to open, and because the sodium then pumped into the neuron's interior creates an electric current in that neuron. And why, then, do axons have sodium channels? Because specialized cells manufacture proteins, forming the walls of these channels, and send them to the axons. And how are the sodium channels situated along the axon? At quite regular intervals.[14] How do they know where to situate themselves? Because other cells, oligodendrocytes, give them orders to do so. How is this information transmitted? By means of a soluble protein. And so on.

The sequence is endless, and so we can compose a new routine, one that is just as exasperating as the first one: "I see a photon . . . photon for retina . . . retinal . . . all-out gallop . . . optic nerve . . . nervous Nellie . . . leeward wind . . . Windemere . . . merely a proton . . . proton on opsin . . . opsin to neuron . . . rondo . . . Dover . . . vertical line . . . line drive . . . drive time . . . time at bat . . . battle ax . . . axon . . . ontology . . . genie in a bottle . . . bottle cap . . . cappuccino . . . nobody . . . body electric . . . electric current . . . rent control . . ."

In short, scientific research is indefinite, an ever-renewed effort to tell the truth about nature.

So, you'll retort, we might as well stop searching, since we will never know the final story and because research is very expensive.

Of course not! Think about everything that scientific research brings us (or brings in for us). The string of questions I just mentioned, which mobilized dozens—hundreds, rather—of researchers over the past half-century, gives us reason to hope for:

- advances in the fight against multiple sclerosis;
- the design of better mechanisms for seawater desalination;
- the construction of chemical computers that are much more effective than current devices and well deserve the affectionate nickname given them by computer scientists in France: *bécanes* [Slang, originally for "bicycles."—Trans.];
- progress in ophthalmology;
- improved understanding of intercellular communication, with the prospect for better controlling it and for applications ranging from embryology to the food industry and cosmetics.

And so on.

But the frustration comes not only from the arduous slowness of science's accession to the (provisional) truth; it is also caused by another constraint, that of constantly having to disturb common sense, not only the common sense of others but one's own as well. To illustrate this, I once again turn to a proverb.

THE SAYING ABOUT THE RED HERRING.   "To draw a red herring across the path" means "to send on a false trail." Most often, one is content to merely say "a red herring," which is allusion enough. This expression originates with the hunt: a red herring, that is, not a live but a salted, smoked, and/or dried herring, by the very fact of its strong odor makes hunting dogs go wrong when it is drawn across the trail they are following: it makes them lose the olfactory trace of the game they are tracking.

The red herring is one of the fundamental narrative structures of the detective novel. For example, Agatha Christie very often devotes each of the chapters in her books to dragging the reader down a new false path, presenting yet another character as a plausible suspect before moving on to another alleged culprit in the next chapter. I find this quasi-mechanical process tedious. But as a metaphor—related to the movement of a fly that bangs itself repeatedly against a window in trying to regain its freedom—and if we grant that the labyrinth is a graphic and symbolic figure, we can see that what we have here is a powerful trope for life.

This trope is also one of the most frequently employed tactical weapons: Louis XI and Bonaparte excelled at its use. Forcing the enemy to rush into a trap is one of the instruments of the art of war, as at Austerlitz, when, once the fog had lifted, the entire enemy army found itself on a frozen lake. Napoleon then began to fire the cannons so as to shatter the ice and drown his enemies. In like manner, the essential principle of Asian martial arts such as judo is to use the impetus, the energy, of one's opponent—by depriving him or her of a point of support in a way as sudden as it is unforeseen—to transform that energy into a sudden fall. For the most formidable red herring is the one that one presents to oneself in rushing to the first explanation that comes along for an event or a phenomenon. Moreover, especially in scientific matters, those hypotheses that appear most farfetched are often the good ones, while what good sense proposes appears afterward as a false path.

THE *SAUGRENU*. "To add one's grain of salt," as the French expression goes, means "to contribute to a discussion, to say or do something original or witty."[15] The sense of this expression has no doubt weakened. Actually, the adjective *saugrenu* (denoting an unexpected event or phenomenon that renders received opinion ridiculous), also composed of *sel* (salt) and *grain* (grain), has retained a stronger sense.

The *saugrenu*—I'll use the word from here on also as a substantive—is a genuine category of understanding, as it is of moral philosophy. The *saugrenu* is not to be confused with the unusual nor with the unexpected. It partakes of the uncanny and is near-synonymous with the ludicrous. The *saugrenu* is the hair in the soup, that which disturbs us right from its surprising and heterodox moment of appearance. Characters such as Ferdinand Lop in the fifties, or Dupont, alias Mouna Aguigui, in the seventies and eighties, were incarnations of the *saugrenu* in Parisian election campaigns. The *saugrenu* is the surreal component of the real. The *saugrenu* shocks, it scandalizes, it disrupts conventions and meanings, it collides but doesn't cause pain: like the jester or the clown, the *saugrenu* remains in the realm of perception, of the intellect, without ever turning into polemic or physical aggression.

The *saugrenu* is a useful—to be precise, even a fundamental—notion in the philosophy of science. At the moment of its emergence, every scientific discovery can be termed *saugrenu*, since it reorganizes first those areas surrounding it and then, little by little, knowledge as a whole, or what we thought we knew before its troubling apparition.[16]

As a matter of fact, I'll borrow from salt—from sodium chloride, to use its technical name—an example of a singular and provocative discovery that, at the time, was experienced as *saugrenu* by some great minds, among them the physicist Arnold Sommerfeld: the Zeeman effect.

What is this effect? Recall that if one casts a bit of salt into a Bunsen burner flame, the flame takes on an intense yellow coloring: the sodium atoms, excited from heating to a high temperature, emit light. These atoms return to their ground state by emitting light. In fact, the light contains two intense lines—called sodium D lines—at wavelengths close to each other that are found in the visible part of the spectrum (produced by making visible light pass through a prism or a grating).

Hendrik Lorentz (1853–1928) had improved Maxwell's electromagnetic theory and provided a new theory of electromagnetic radiation. As a result, in 1896 Pieter Zeeman (1865–1943) discovered that sodium lines, when the light emitter is placed between the poles of a magnet (i.e., when sodium atoms returning to their ground state are subjected to an intense magnetic field), split into at least three components. This Zeeman effect had been predicted by Lorentz; the two

Dutch physicists consequently shared the Nobel Prize in physics in 1902: theory makes common cause with experiment, not only to lend it meaning or to express it in mathematical equations but also to describe possible ways in which reality might behave, which it remains for experiment to find out.

How can it be that the Zeeman effect was not discovered earlier? Actually, Michael Faraday had conducted the same experiment as early as 1862 but had failed to find this phenomenon. Zeeman considered Faraday as the greatest experimental genius of all time; he surmised that if Faraday had tried the experiment, then it was certainly worth repeating it, and this was how he came to his discovery.

Before getting to the meaning of the Zeeman effect for the atom and to what about it is (or was) *saugrenu* at the time of its discovery, let me mention one of its very first applications: light reaching us from the sun has a continuous spectrum, which shows the lines of absorption by hydrogen, magnesium, iron, and (you guessed it) sodium atoms. Moreover, sunspots show a Zeeman effect, which proves that they are the site of intense magnetic fields. Astrophysicists would later discover a Zeeman effect in stars other than the sun.

Arnold Sommerfeld explained the Zeeman effect in 1916: every electronic transition takes place between two energy states of the atom, which are like the floors in a building.[17] When they are excited by an increase in temperature, the sodium atoms are lifted up to the higher floors; in other words, the electrons in these atoms are made to reside in higher energy states. They spontaneously move back down to the lower floors—to lower energy states—by emitting a radiation, in this instance, the sodium D lines.

The Zeeman effect is the observation of an increased number of lines when the atom is subjected to a magnetic field. Sommerfeld explained it, since every magnetic field is endowed with an axis (the one that runs from the magnet's north pole to its south pole), by postulating that the presence of this axis serves to orient space around it, and the orbits around which any given electron moves are themselves also oriented in relation to this axis. Not all such alignments are permitted. Only certain alignments in limited numbers (four for an electron in a sodium atom) can be observed: this is what is called spatial quantification. It was to receive its full explanation only with the discovery of the electron spin—a property shared by other elementary particles—following another surprising effect (and one that is likewise a priori *saugrenu*) called the Stern-Gerlach experiment, dating from 1921–1922, which I will not discuss here. A crude and inadequate—because on our scale—image of electron spin is the rotation of a top.

Until this point, I have discussed only absorptions and emissions (or spectral transitions) resulting from *electrons* in atoms residing in different energy states. Other transitions stem from the energy states of atomic *nuclei* when atoms are placed in a magnetic field. This is also a Zeeman effect.

The nucleus of the sodium-23 atom (this is the stable isotope, natural abundance 100 percent) thus gives rise to four distinct energy states, or Zeeman levels. It is possible to induce transitions between these levels by submitting the sodium atoms to radio waves, in the range of several tens of megahertz, just like FM radio; this is what is called nuclear magnetic resonance, or sodium NMR, a tool put to frequent use in my own research. This is not the place to present the details, but this tool has allowed us, among other things, to time the very rapid chemical reactions undergone by biomolecules, as when sodium Na^+ cations pass through a cellular membrane in channels made of a molecule called gramicidin, and to discover the self-assembly of guanosine, one of the four kinds of units (or bases) carried by DNA (this self-assembly is put to good use by nature for pairing chromosomes during the phase of cell division called meiosis).

Thus it was with humble kitchen salt that electron spin was discovered. This episode in the history of science contains a number of layered meanings:

- the *cumulative character* of the story: Zeeman built on foundations laid by Faraday, and Sommerfeld in turn came up with his theory to account for the observations by Zeeman;
- the *uniformity principle* of physical laws, whence flows the discovery, at a distance (quite a distance!), of the magnetic field associated with sunspots;
- *change of scale*, in going from the macroscopic to the microscopic level, leading to the appearance of brand-new phenomena such as the quantification of space or the presence of spin in elementary particles such as electrons or nuclei.

How can one give the *saugrenu* its chance when managing scientific research? The best answer, plainly, is to trust in the singular, extraordinary individual, whether he or she works alone or as a group leader. Such people won't let themselves be swayed by conventional wisdom. Neither will they be swept away by scientific fashions when many of their peers rush after a glittering topic promising the twin rewards of glory and money. As it turns out, most such lemming-like vogues are short-lived and almost always mistaken.[18] Marginality makes the pioneer as deviation makes invention (consider, for example, the Macintosh computer, initially conceived as a children's toy, or the Post-It, used first as a bookmark in a missal). And if marginality makes the pioneer, catching a stereotype by the tail can make for good ad copy.

PUNNING IN THE RAIN. "When it rains, it pours." This motto, along jointly with the logo of a little girl sheltering under an umbrella, identifies the Morton Salt company. I wish to comment here on the textual part of this emblem, on its *suscriptio*, as this was named in Latin in northern Italy at the time of the Renaissance, when Alciato, then a law professor at the university in Pavia, published the first collection of emblems.

The Morton Salt motto is easily understood in its literal meaning: "even in rainy weather, one does not have any trouble with this salt; it flows as usual, and one can even pour it with no difficulty." The mere length of this explication shows that the Morton Salt slogan draws the sharpness of its impact from its concision. Its writer was able to condense in this pithy formula the asset the company meant to become mentally conjoined with its product.

Another aspect of the memorable Morton formula is the fascinating ambiguity, the pun on the pronoun *it*. The motto is shaped like a logical proposition of the form "if A, then B," where A equals "it rains" and B equals "it pours." But the form of the statement is deceptive, and in a most interesting manner: hence the feeling of an instantaneous frustration followed by quick resolution in a chuckle, in a sort of inner smile, as soon as the mind has found the trivially easy solution to the dilemma before it. The problem arises because the grammar of the language encourages the reader to identify the indefinite pronoun *it* in the two halves of the motto with a single subject or referent. Thus the first spontaneous reading is meteorological. It might be paraphrased as: "When there is rainy weather, this rain degenerates into a downpour, as during a tropical storm."

That this statement is unintended is clear, however, because it leaves out its referent, the object of the description, namely, a container filled with salt. Hence the mind pushes aside the trope it had first grappled with, in the shape of a rhetorical exaggeration (the figure known as the hyperbole), "when it rains, it pours," and allows for the likelihood of a pun instead. The mind opens itself to another meaning, the double entendre understanding of the motto. This second viewing of the sentence enforces a dissociation between both occurrences of the pronoun *it*. When the little word *it* shows up, its second appearance does not refer to the same entity as the first. In persona one, "it" denotes the weather; as persona two, "it" denotes Morton Salt.

The verbal ambiguity helps to make the statement memorable. It shares such a structure with quite a few proverbial sayings. An empirical observation (in this case, "when it rains") is reiterated but the carbon copy ("it pours") is not superimposable on the original. And indeed the Morton Salt motto did originate in a proverb "it never rains, but it pours" shaped as a hyperbole. "I don't like you, I adore you" has an identical form.

At the turn of the twentieth century, salt was shipped in the United States by the Morton Salt Company—which Joy Morton, son of Grover Cleveland's secretary of agriculture, had started in 1886—in small cloth bags. Moisture made it harden like rock. The client had to pound the bag with a mallet in order to pulverize the contents for storage in a jar on the kitchen shelf. By 1907 Morton engineers were grappling with the problem. It being salt, they licked it; after four years of research and somewhat raw tongues, they came up with an astute solution: addition of a tiny amount of magnesium carbonate to the sodium chloride absorbed any moisture and made it possible for the salt to pour out of a round cardboard container equipped with a metallic pouring spout.

Morton Salt then hired the N. W. Ayer and Company advertising agency to come up with a logo and with a motto. The ad, consisting of the little girl with an umbrella in one hand and Morton's cylindrical salt container under the other arm, first appeared, together with the copy I have just commented on, in *Good Housekeeping* in 1914.

The year 1914 also saw the start of World War I, when it poured on soldiers not only rain and snow but shrapnel, shells, and bombs. I cannot help being reminded of the carnage by the very innocence of the Morton Salt motto.

FROM SALT TO SALTS.   The part of my high school chemistry course devoted to sodium chloride and its properties began with a generalization: just as sodium chloride, with the formula NaCl, is the result of joining caustic soda, NaOH, and hydrochloric acid, HCl, with the attendant production of water H_2O, so the combination of any acid and base whatsoever forms a salt plus water. We were given as examples salts such as potassium chloride, silver nitrate, and barium sulfate. And the chemical equation "acid + base ⟶ salt + water" was presented to us as a law of nature, inscribed across the pediment of chemistry. Do you want to know whether an unknown substance is an acid? Put it in the presence of caustic soda: this reagent will act as a touchstone, so to speak. If the reagent forms a salt and water, one can be certain that it was an acid.

As bad luck would have it, aside from water, the three other terms of the aforementioned master equation are undefined, or if they are defined, it is in a circular fashion. What is an acid? It is a substance that, placed in the presence of a base, yields a salt and water. What is a base? It is a substance that, placed in the presence of an acid, yields a salt and water. And what is a salt? It is the result of the union of a base with an acid.

Further along in their apprenticeship, those chemists who had not yet been discouraged by these odd features of their training advanced to another definition of a salt, this one also an operational definition. Salts dissolved in water—incidentally, in the case of an insoluble salt like barium sulfate, this is a thought experiment rather than an empirical observation—are electrolytes: such aqueous solutions conduct electric current. Unfortunately, this improved definition remains unsatisfactory, because the category of electrolytes, though it does indeed comprise salts as a subset, also includes other compounds (among them acids and bases).

Those of us who stuck with it learned, still further along in our studies, that there exist other kinds of acids and bases than those forming a salt and water when reacted together. And, conversely, we were informed of the existence of acids and bases, organic and not mineral, whose combination formed, in addition to water, esters (which are not electrolytes) rather than salts.

The first vulgate was incomplete. Those that followed relativized it by inserting it into a more general framework, and it returned to the status of a particular case. We then learned, during the course of doctoral studies at the time, that any chemical compound could be looked at as an acid, a base, or even a salt. Better yet, one could turn every chemical compound into an acid or a base by adding or subtracting a particle such as a proton.

In short, our apprenticeship in chemistry had followed a course of abstraction. Taking off from a positivist and phenomenological introductory education, at the level of evident material properties, we gradually had acceded to an advanced formalist and deductive education. Along the way, chemistry had vanished. My personal experience conforms with that of many generations of students. And, in my view, this is not the way to proceed.

In fact, chemistry resides neither in particularism and substantialism (descriptive monographs on substances such as NaCl) nor in the generality of supposed laws, which seek to mimic the laws of physics, such as the laws of conservation or Newton's laws. It is clear that education in chemistry from the start (and not progressively) must be both conceptual and experimental (stressing observation and its interpretation). The mistake in the education I received came from locating the conceptual element in the wrong place. The notion important to convey is not that of substance—acid or base—but the notion of the relation between them, which, in fact, defines them both, namely, the notion of the acid–base pair. In short, we were all subjected to an Aristotelian education in chemistry when a Kantian education was needed: it is not the entities that matter but only their interactions. Chemistry is the science of such interactions.

One can attribute this glaring error in instruction to the historical significance of the notion that was taught (but it would have been so much better if we had been made aware of this and if it had been explained to us): salt was of primary importance to alchemists. Furthermore, dualism has been so fundamental, for chemistry as for alchemy before it, that this historical heritage came to color and taint education in a lasting manner.

seven **myths**

What differentiates science from myth? The first offends common sense, while the second readily exalts it. Various mythologies raise salt from an object of primal necessity to the rank of an existential virtue, celebrated and sanctified.

The Bible enjoins ritual and liturgical use of salt. The Romans also had a specific festival in which Salian priests danced, chanting a psalm so ancient that it had become incomprehensible. As for the Aztecs, they paid homage to a salt goddess in a ritual that culminated in the sacrifice of a young man who had reigned as prince of pleasures for a solid year.

In numerous cultures, defining a human sphere from which evil is excluded is accomplished ritually by casting salt. The saltcellar on the dinner table furnished the means for this purification in addition to filling a more practical function. Food touched by salt in this way is removed from the contingencies of scarcity and illness.

For centuries on end, saltcellars, at least among the nobility, were symbols of grandeur and luxury (evidence of this is the opulent piece of silverwork that

Benvenuto Cellini fashioned for François I). At the heart of the table's finery, they brought together decorative and sacred art, the meal and the Last Supper.

From the family meal to the feast among friends, salt helped people become aware of life's meaning. All these offerings represent a carpe diem, a consciousness of the preciousness of salt, to be used with care as great as that taken in raising a child—and the other way around as well.

Still other related metaphorical allusions can be mentioned, like the Stendhalian notion of crystallization and, for Ramakrishna, the doll of salt dissolving in the ocean, after the image of humankind melting back into God.

RITUAL AND LITURGICAL USES OF SALT IN THE BIBLE.   Since we belong to a Judeo-Christian culture, any examination of rituals linked to salt must return to the Bible. For the Jewish people, salt is inseparable from religion. The Bible abounds in injunctions such as "with all your offerings you shall offer salt" (Lev. 2:13), the meaning of which is clarified by the accompanying prohibitions against offerings of honey or leavened bread.[1] Honey and leavening in fact represent foods that are subject to fermentation and thus to deterioration and spoilage. Salt, on the other hand, retains its nature in a dry climate like that of Palestine. This desirable integrity would symbolize the covenant entered into between humankind and God as the "covenant of salt for ever" (Numbers 18:19).[2]

Hence the Temple of Jerusalem included a slaughterhouse, to accommodate the sacrifice of cattle according to the ritual prescriptions, and a hydraulic mechanism,[3] on the one hand, for the washing and salting of this meat, using a brine, and, on the other hand, so that the faithful might wash themselves and so be worthy of this sacred place. The courtyard of the Temple of Solomon thus had two thousand baths, fed by the Gihon Spring, whose waters flowed through the temple.

Salt was renowned for its purifying and antiseptic properties. It served to purify the unhealthy waters of Jericho (2 Kings, 2:20–21). Newborns were rubbed with salt (Ezekiel 16:4).[4] Of course, salt also stood for flavor in food,[5] and the Jewish people adopted the practice of putting salt on the bread used to bless a meal before it begins.[6]

The other great cultures of classical times also accorded salt a privileged place. We retain a trace of this, linked to an odd festival in Roman mythology.

SALT AND DANCE.   The Greeks had many names for the sea. *Hals* was one of these, the one that was worth its salt. We can track this word to the salt mines: even today, in Austria and Germany, towns are called Halstadt or Halle.[7]

Hals, the salt sea! The root of this term is found in abundance in Greek mythology, curiously, in the dispute between the two violent gods, Poseidon, god of the ocean, and Ares, god of war.

These tales of southern peoples are, of course, family stories: the Aloades, sons of Poseidon, as one can guess from their names,[8] enclose Ares in chains for thirteen months in a brazen jar. Halmos is a character whose two daughters marry Poseidon and Ares. Halia or Halimede was one of the Nereids, daughters of the Old Man of the Sea.[9]

But another homonymous Halia married Poseidon. Their six sons, inflamed by Aphrodite, who was doubtless jealous, raped their own mother. Poseidon, incensed, banished them to the underworld with one blow of his trident. Halia, in despair, flung herself into the deep.

Halirrhotios, another of Poseidon's sons, also hot-blooded—his name literally means "roaring salt sea"—attempted to rape a daughter of Ares, who killed him. But Poseidon convened a tribunal of the gods, which tried his colleague on a hill that thereafter was known by the name of Areopagus.

As we know, Roman mythology took up and translated a good part of Greek mythology: Ares became Mars, and Poseidon became Neptune. As for Hals, the Romans called her Salacia, the salt one; for them, she was the allegorical personification of salt water. In the transition from Greek to Latin, the root *als* had become *sal* through a circular permutation of the three letters. This occurs rather frequently when one language borrows from another or from itself. To cite an example of such a doublet that is also the product of a permutation, the same French river is called the Olt at one point and the Lot at another.

The Romans also copied without making modifications. As a part of their ancient history, they told the story of a certain Halesius, a son of Neptune, who had come to Italy. Morrius, king of Veio, had even instituted a rite in his honor, that of the *carmen saliare*, a song that the priests of Mars voiced while dancing in two annual ceremonies, one of which was held in March (at the very beginning of the Roman year) and the other in October. By that point, the language had become so archaic that no one understood the song's lyrics any longer.

There were twelve of these priests of Mars. Why twelve? Numa, the second king of Rome, had created their collegium to commemorate the granting to Rome of a protective shield, or *ancilia*, fallen from the heavens when the city was suffering from an epidemic. Numa had twelve replicas of this *ancilia* made by an artisan, Mamurius Veturius, whose name meant "old man Mars." The priests in the service of Mars, or *Salii* (Salian priests),[10] each looked after one of the twelve *ancilia*. On the occasion of the new year, these Salian priests paraded, singing their incomprehensible litany, but they also chased away a person dressed as a little old man—just like our Santa Claus—who symbolized Mamurius Veturius. It was a propitiatory rite: the twelve *ancilia* symbolized the twelve months of the year, and the expulsion of the old man signified that the year gone by was being buried. When night fell, a feast was held.[11]

But why this name, Salian priests? Classical authors differ on this point. Some hold that the term derives from Salos, a comrade of Aeneas. As for Ovid, he links the Salians' name to their dancing parade, since the verb *salire* means "to leap," "to dance." Other authors see the term as derived from Salios, a small sister tribe of the Latinos.

The ceremony of the Salian priests, at the beginning of March, is one of many variants of Indo-European legends linked to the tripartite division of society into priests, warriors, and peasants. I am tempted to connect the *Salii* with some of the other legends mentioned and to make them the symbolic guarantors of Rome's well-being and of a peaceful year, after the god Mars had been contented with the ceremony in his honor and was kept calm (as when the Aloades had put him in a cauldron for a whole year). Some, from Ovid to contemporaries such as Lewis Thomas, make the conjecture of a common etymology for *salt* and *salire*, for the noun *salt* and the verb *to dance*. Perhaps because *Hals*, the salt sea, undulates in the wind?[12]

If we turn our focus to Amerindian culture, which differs so completely from our Greco-Hebraic sources, we find a mythical role there for salt as well, one not so far removed from its role in the Roman festival just described.

AZTEC BACCHUS.   There are many cultures that put a young man to death after having treated him like a god for an entire year. The Aztecs, after having chosen the future victim, urged him to behave like the ultimate dandy, to delight in all life's pleasures, though nonchalantly, and to go about fluting a little tune to announce his arrival. The fine youth wed four girls before his sacrifice to Tezcatlipoca, the sun god. They represented the four young goddesses Chalchiuhtlicue (goddess of fresh waters), Xilonen (goddess of young maize), Xochiquetzal (goddess of flowers), and Huixtochiuatl (goddess of salt and dissolute life).

The Aztecs practiced human sacrifice in a regular, institutionalized, and nearly humdrum manner. These sacrifices were supposed to repay a debt to the gods, as well as to maintain the regularity of planetary movements. The original myth of the creation of humankind by Quetzalcoatl held that a shedding of the god's blood gave birth to humans; it was thus appropriate that this blood debt should be repaid. In addition, after his birth the sun god, Uitzilopochtli, had to do battle with his sister, the moon, and with his brothers, the stars. Armed with the serpent of fire, he was victorious, and his win secured for humankind one more day of life. Therefore it was right to give him thanks for this good deed by offering him the food he was fond of, *chalchiuatl*, that is, human blood. In a way, *chalchiuatl* was the fuel that ensures the smooth functioning of the moving solar body. As Jacques Soustelle described it, "The regular return of the Sun is indispensable to the salvation of the universe. The Sun needs blood to give him energy."[13]

Considered in another way, Uitzilopochtli symbolized the resurrection of the warrior killed in battle: reversing this symbolic meaning entailed sacrificing to him a warrior taken prisoner during combat in a war whose aim was to capture future sacrificial victims.

The captive warrior was chosen for his physical resemblance to a young god, smooth-skinned, physically flawless, a young Apollo who would subsequently be put to death so that the sun would continue on its course. For a solid year, he enjoyed the life of a prince. Fitted out in beautiful dress, so as to pass for the god Tezcatlipoca, a brother of Uitzilopochtli and Quetzalcoatl, he strolled the streets, smoking his gilded bamboo pipes, a bouquet of flowers in hand, or playing on baked clay flutes. For a whole year, the handsome dandy lived the high life, enjoying a princely existence and, like a king of the carnival, living a life of pleasures and privilege.

Then came the day for him to be put to death, the Toxcatl ceremony in the sixth month of the Aztec calendar. On this day, festivities, ceremonies, and feasts were held in honor of the young man. The entire populace participated in them. Then, accompanied by his four wives and his retinue, he was led to a small temple beside a lake.

His entourage left him. Only a few pages remained. At the foot of the temple, his pages also departed, leaving him alone with his pipes and flutes. He then mounted the steps leading to the temple, breaking his flutes as he climbed. At the top were the priests. Four of them stripped him of his fine clothes and held him stretched out atop the sacrificial stone, while a fifth plunged an obsidian knife into his chest and tore out his heart.

The young man personified Tezcatlipoca, one of the gods at the peak of the Aztec pantheon in the period from the fourteenth to the sixteenth century. God of the night sky and the Great Bear constellation, the avenger god, all human actions, good or bad, were within his ken, and he rewarded them with either honors and wealth or poverty and illness. The Toxcatl sacrifice in his honor also had a moral, applicable to all: it was to recall to mind the fleeting nature of existence, even for the privileged few able to enjoy all life's pleasures.

A word remains to be said about the protective goddesses of the four wives, for a year, of the sacrificed young warrior. Xochiquetzal was the Aztec Venus, goddess of beauty, of sexual love, and of the domestic arts. Associated with plants and flowers, she was a divinity of flora, reputed to come from Tamoanchan, the luxuriant paradise of the West.

This goddess of love was at times identified with the second protective divinity of the young wives, Chalchiuhtlicue, the goddess of fresh waters and fertility: in the cosmogony, she had been the fourth of the preceding suns. And it was under her reign that humans had first tasted maize.

In fact, the third divinity, Xilonen, in whose name one of the wives was also chosen, was goddess of young maize, and can also be seen as a guarantor of fertility and of abundant harvests.

The fourth goddess was Huixtochiuatl, whose name meant Lady of Salt and who symbolized at once salt water, the saltworkers' guild, courtesans, and dissolute women.

Life is a fleeting moment to be savored.

THE PROVERB OF THE ASPERSION. This proverb is common to many cultures. *Shio* means "salt" in Japanese; *maku* is the verb meaning "to scatter, to sprinkle, to throw." The meaning of the gesture is the summoning of the sacred. The space in or on which salt is cast is sanctified. It is the gesture, the sacramental rite, that one sees sumo wrestlers make before a bout to purify the ring.

The linking of salt with religious ritual in both Japanese civilization and Judeo–Christian culture can be accounted for with the same explanation: it symbolizes immutability. It is a food, or an accompaniment to food, that is incorruptible, that stands for invariance and permanence and thus can be taken for a feature of the divine.

In numerous cultures, pouring salt also has the superstitious virtue of protecting one from evil or from the devil. In Japan, when guests have been unpleasant, one pours out a bit of salt after their departure to cleanse the premises. In Scandinavia, one sprinkles a pinch of salt about to protect oneself from evil spirits and demons. In Sweden, one adds a bit of salt to the milk to protect the precious cow. Again in Sweden, as in ancient Rome, spilling salt is considered a bad omen.

SALTCELLARS. In the Middle Ages, these salt containers for salt had great symbolic significance. Never that small, their size indicated the wealth of their owners. Their distinction derived first of all from their rare and precious contents. For this reason, saltcellars became elaborately worked silver pieces, ornamental and symbolic, greatly varied in form, even combining several different functions.[14]

Even the seating arrangement of guests at the table and their relation to the location of the saltcellar took on a meaning that was ritualized and became protocol. The saltcellar, occupying the center of the table, defined a line of demarcation between the distinguished guests, those that found themselves "above the salt," and the others— regulars, poor relations, inferiors—seated "below the salt." Today, the English language continues to retain this social distinction (which, though no longer marked by the saltcellar, is no less lasting).[15]

BENVENUTO CELLINI. In the Middle Ages and the Renaissance, the saltcellar was a genuine jewel, since it contained a good, salt, that was considered rare and precious. On a lord's table, the saltcellar was, with the *nef* [A dish made of gilded silver in the shape of a ship, used to hold the lord's cutlery and napkins.—Trans.], the most splendid object, so precious were its contents; it was often even put under lock and key until the sixteenth century.

That century saw the development of the most illustrious saltcellar in history, commissioned by François I from Benvenuto Cellini (it can be seen today in the Imperial Treasury of Vienna). The Florentine artist had merely to pull out a drawing already present in his portfolio when the French king expressed his desire for a saltcellar to accompany a basin and ewer he already owned. Excited by Cellini's design, François I immediately had him presented with one thousand gold crowns to create this saltcellar made of gold and enamel.

Here is the first of Cellini's descriptions of the saltcellar, at a point when it was only in the design stage. It clarifies one aim of his allegory, in which salt is construed as a product of the fertile union of earth and sea:

> I made an oval shape, a good deal more than half a cubit in size—in fact almost two-thirds—and on it I modelled two large figures, to represent the Sea embracing the Land. They were a good deal more than a palm in size, sitting with their legs entwined in the same way as certain long branches of the sea cut into the land. . . . Underneath him I positioned the four sea-horses, and I placed the trident in his right hand. I represented the Land by a woman, whose beauty of form was such that it demanded all my skill and knowledge. She was beautiful and graceful, and by her hand I placed a richly ornamented temple; it was placed in the ground, and she rested her hand on it. I made this to hold the pepper. . . . Beneath this goddess, on the part which was meant to be the earth, I grouped all those wonderfully beautiful animals that the earth produces. Corresponding to this, I fashioned for the Sea every kind of beautiful fish and shell that the small space could contain.[16]

The finished piece remains close in execution to this initial plan.

Would peasant saltcellars be any less beautiful? Is the Renaissance artist far removed from the traditional craftsmanship we admire even today in some rustic Alpine saltcellars?

DECORATIVE ARTS: FROM COLBERT TO THE QUEYRAS. A blond wooden cylinder, with a radiating geometric figure representing the sun carved on the top; not very heavy, it has been carved out, which one notices after having released the clasp, made in the same larch wood as the rest of the container, and having swiveled the cover around on its peg: it is a saltcellar from the Queyras similar to those that shepherds used to take to the mountain pastures, whittling new ones during their long stays in the high meadows there.

Today, tourists buy these superb objects, the products of a popular craft industry. Their beauty is one with their function: to keep salt dry. They are so well crafted that once the box is shut it is hermetically sealed.

For that matter, this could perhaps serve as a differentiating characteristic: where the bourgeoisie's utilitarian objects receive an embellishment, a decorative detail, popular art, for its part, gives the impression of a deep harmony between form and function. Granted, the Queyras saltcellar is decorated with a simple sun,[17] but this geometric, stylized decoration—a decoration that, by the way, appears on other wooden objects crafted in Queyras and so indicates the object's origin—remains discreet and serves to identify the object's provenance rather than as ostentatious display.

This is the moment to recall another difference that for centuries on end has been superimposed on the one just cited: in the kitchen, we find boxes of salt that are rather bulky, crude, and sometimes very lovely, as if by chance; on the dining room table in a bourgeois home, especially for guests, we find smaller, more highly crafted saltcellars, at times miniature masterpieces of silverwork, for a wealthier look.

In the sixteenth century, the saltcellar becomes a bit more commonplace, often being made of enameled metal or ceramics.[18] In the seventeenth century, it is included in the dinner service. In the eighteenth century, the style in fashion is silver saltcellars, often including two blue crystal salt dishes.

Such saltcellars for dining are still specimens of decorative art today. Let me venture to define decorative art: it makes functional objects attractive. Let me add another definition: if bourgeois luxury goes hand in hand with a certain ostentation—showing off one's wealth without the slightest timidity—decorative arts supply such luxury with beautiful objects, most often produced in small quantities by artisanal methods, thus remaining relatively accessible, in contrast to the unique, singular work of art that, for its part, may be extravagantly expensive.

In France, we owe to Colbert the concrete embodiment of this paradox of inexpensive art. He recognized, approved of, and encouraged the need for the new bourgeois class, which had come to share power with the aristocracy, to enjoy a nascent respectability and access to reasonably priced luxury objects that manufactures could produce. Colbert saw in this another advantage, namely, that of protecting and so preserving the precious practical knowledge held by glass, leather, weaving, and furniture artisans. This is how Diderot described it:

> Colbert considered the industry of the people and the founding of manufactures the most reliable resource of a kingdom. In the opinion of those who today can discern true worth, the state benefited no less from a man who filled France with engravers, painters, sculptors, and artists of all types, who wrested from the English the secret of the machine for producing hosiery, from the Genoese their velvet, from the Venetians their mirrors, than it benefited from those who vanquished the enemies of France and took their fortresses. In the eyes of a philosopher a sovereign may deserve more praise if he has encouraged men like Le Brun, Le Sueur, and Audran, if he has had the battles of Alexander painted and engraved, and the victories of our generals represented in tapestry, than he would for having gained those victories.[19]

Alas, came the industrial revolution: one example suffices to evoke the ruin, with few exceptions (Gobelins, Sèvres), of the applied art manufactories established by Colbert: the French production of Indian-patterned cotton goods, set up by Colbert to break the British East India Company's monopoly, fell victim, in the second half of the nineteenth century, to industrial imitations and synthetic chemical dyes that were overly bright and lacked the subtle nuances of the natural coloring agents (madder root, woad, walnut stain) used hitherto.

In the meantime, the industrial revolution had made salt and saltcellars totally ordinary. In the case of salt, its use as a raw material for the nascent chemical industry had sharply increased its production—especially through the mining of the many salt domes that crop out more or less regularly across the great central plain of Europe—and revolutionized its status. As for the saltcellar, demotion was due to mechanization, which allowed production of one identical model on a very large scale at a cost of a few pennies a piece.

But some artists rebelled against the leveling effect of the low quality of these industrial products relative to the products of the craft industry. The first to do so was the very remarkable William Morris, in England, with his Arts and Crafts movement, who extolled a return to a tradition of artisanal masterpieces and drew his models from the Middle Ages, both in the matter of technical perfection (embroidery, tapestry work, printing on cloth or paper) and in the decorative patterns he drafted. Prefiguring such mass-marketing successes of today as the Elle boutiques or the Ikea stores, William Morris aspired to give every moment of daily life its dignity and beauty, something that could be conceived, in his view, only from the egalitarian perspective of socialism.[20]

Closer to our time, Paris in the 1980s saw another of these periodic rebellions of artists against mechanization and against what they experience as the banality of daily life, attributable to a soulless industrialization. Interior design artists, of whom Philippe Starck is the leading light, seek to situate their creations between an art of luxury, reserved for a privileged few, and the industrial design of mass-produced products.[21]

Saltcellars, at the start of the third millennium, thus belong to three categories: wedding lists, and these are luxurious objects, sometimes in leaded glass, created by manufacturers such as Baccarat, Christophle, Corning, or Daum; everyday industrial products, sometimes extremely handsome, in steel, glass, or even plastic; and, as a third category, trying to create an intermediate place for itself, artisanal products in wood, earthenware, stoneware (for a long time, throughout the sixties and seventies, the Ratilly saltcellar was the mark of elegance on a French table),[22] or glass.

I dream of an exhibit—or a series of television broadcasts—about the saltcellar of François I. When was it shown in France last? Of course, the various accounts of its creation by Benvenuto Cellini would be well placed there. The saltcellar would be accompanied by artworks (or reproductions of them) that shared its theme, the meeting of Aphrodite and Poseidon: to start with, the four figures on the base, representing the parts of the day and inspired by Michelangelo's sculptures for the Medici chapel in Florence (*Day*, *Night*, *Dawn*, and *Dusk*). One could thus evoke the notion of antiquity, as it occurred at a certain point in the Renaissance, through Cellini's masterpiece.

This exhibition on the theme of salt containers could open with all that the history of ideas can teach us about daily life, about private hospitality, travel, and celebrations through centuries past. It would provide information on the saltworks, and the craft of the saltworker, through technical manuals and travel books from the Renaissance forward, as well as photographic material on what remains of these in our time. It would show the clothing and tools used by the saltworkers, explaining their functions and names: *batoué*, *boïette*, *las*, *palle*, *salgaïs*, or *tranche*. Another section would treat the salt mines. A fourth section would be devoted to igneous salt and to the Chaux saltworks built by Claude-Nicolas Ledoux.

One would further make every effort to answer such questions as To whom was the saltcellar bequeathed in a will? Would salt be carried on trips, say, in the sixteenth century? Would the saltcellar be a part of a bride's trousseau? Would it appear on inventory reports after a burglary?

Thus, in two parallel sections, one would work to present a historical account—covering roughly the last eight centuries—of hospitality and friendship, of cooking and of dining customs, using salt chests from the hearth;[23] kitchen salt boxes; table saltcellars of varying elegance, size, and ornateness; saltcellars that once belonged to famous figures; hotel and restaurant saltcellars; saltcellars for use outdoors or while traveling and, later, while camping. Impressive medieval saltcellars would recall the role played by this object in establishing the ranking of guests at a lord's table.

One would also attempt to connect the styles and constitutive materials of the saltcellars with those of other objects from private domestic life: thimbles, scissors, candlesticks and sconces, tumblers, silver, and so on. Efforts would be made to draw comparisons between nations, between the lands of the great gabelle and the lesser gabelle, between cities and rural areas, between civil and military saltcellars . . .

The exhibition as imagined would have a didactic aim; this, then, is an appropriate time to mention instruction, which also has a link to salt, at least in a Japanese saying.

THE SAYING ON THE PINCH OF SALT. "Teshio ne kakeru": in this Japanese saying, *teshio* is a small cup in which to serve salt, and *kakeru* is a verb meaning the action of pouring. Therefore the meaning of this expression is "to help oneself to salt on one's own." The emphasis is on this reflexive aspect, on the autonomy of the gesture of moving one's arm forward to take hold of the dish, from which one is about to take a pinch of salt.

The figurative sense of this expression is "to raise a child in taking great care of him or her." The transposition of the concrete image of salt taken at the table to a principle of childrearing takes place in two moves: first, the metaphor with which the young human being is likened to a rare and precious good that imparts flavor to foods and, consequently, to life; and, second, the attitude of the hand and fingers at the moment they take up a bit of salt, which acquires the symbolic meaning of a gesture to protect something eminently fragile, something very vulnerable, that must be watched over so as to protect it as well as possible.

This saying has profound meaning. Salt is not a food like any other: it is incorruptible. A child is also incapable of decay. Though he is weak, he nonetheless has a set personality that deserves respect, if only for the values he inherits from his ancestors and then himself transmits.

Childrearing in the Far East, particularly in Japan, bears the stamp of the civilization of farmers in which it evolved. Without wishing to posit too stark a contrast, in Western nations we raise our children (in itself a very telling expression) like small domestic animals insofar as ours is a civilization based on livestock raising. In Japanese culture, the child represents instead both the promise of a future contribution to the physical efforts of the human group, which, through assiduous daily labor must look after the rice paddy to ensure its productivity, and a good as precious as the annual rice harvest (or the annual fish take) that guarantees the group's survival. Thus the image of salt in this saying carries the double meaning of a food, on the one hand, and of a precious, nonrenewable good insusceptible to spoilage, on the other.

I now move from this notion of salt as incorruptible to its transmutation, as if by magic, into an emblem of amorous desire, which is thereby made lasting: salt's crystallization, as Stendhal saw it and wrote about it, immortalizes an image that he saw in passing but that made a lasting impression on him.

STENDHALIAN CRYSTALLIZATION.   In *De l'Amour* (*On Love*), Stendhal lent to the term *crystallization* an enduring symbolical meaning: "Under pain of appearing unintelligible from the very beginning, I thrust on the public the new word *crystallization*, coined to express vividly that mass of strange fancies which one imagines to be true and even indisputable facts in connection with the person one loves."[24]

Here is the definition Stendhal gives it:

> In the salt mines of Salzburg a bough stripped of its leaves by winter is thrown into the depths of the disused workings; two or three months later it is pulled out again, covered with brilliant crystals: even the tiniest twigs, no bigger than a titmouse's claw, are spangled with a vast number of shimmering, glittering diamonds, so that the original bough is no longer recognizable.
>
> I call *crystallization* that process of the mind which discovers fresh perfections in its beloved at every turn of events. (6)

The scientific tone is somewhat surprising in a book on love. But it is necessary to recall the passion Henri Brulard (that is, Stendhal) felt for mathematics, a subject for which he used the same terms (love, cascade, enthusiasm, adoration, passion) as he did for the passion of love.[25] For Henri Brulard, love of mathematics prepared him for love of women: it preceded it in time, but he saw it also—a projection into the future—as a means of leaving Grenoble, of hoisting himself up into Paris society (which the young Beyle did, having been admitted to the École polytechnique, which he then declined to enter).

Stendhal wanted his treatise *De l'Amour*, which he meant as "an exact and scientific description,"[26] to have the power to convince, with the rigor, thus, of a mathematical proof: "This book explains simply, rationally, mathematically, as it were, the different emotions which follow one after the other and which taken all together are called the passion of Love" (xv). What is more, the author sets out this "passion of Love" as if it were a geometric figure of sorts: "Imagine a moderately complicated geometrical figure traced with white chalk on a blackboard: well! I am going to explain this geometrical figure" (xv).

Couldn't one say that a crystal, with its "multitude of different shapes, at once regular and sharply pronounced,"[27] with its triangular, pentagonal, or otherwise shaped facets, is an excellent concrete basis for such a geometric figure? Especially since the brilliance of a crystal discovered in the subterranean night can be compared to white chalk on a black slate?

Certain statements in *De l'Amour* return to mathematics to extend the phenomenon of crystallization to that subject: "There is even crystallization in mathematics (e.g., the Newtonians in 1740) in the minds of those who cannot visualize at the same moment all the steps of the proofs in which they believe" (18; translation modified). Or, to borrow the terminology of algebraic equations: "so long as you are on a distant footing with the person you love, crystallization takes place as an *imaginary solution*" (17). Indeed, to Stendhal, love and mathematics share that "in this frightful passion, *a thing imagined is always a thing existent*" (113).

In addition, doesn't mathematics excel in what seems to be "a necessary condition for happiness through love" (108)? "Mathematics has the charm of saying only things that are certain, only the truth, and almost the whole truth."

Truth is clear and pure like a crystal. The crystal was even a standard poetic trope, used in particular to describe a totally limpid body of water; numerous examples of this could be cited. But the poet's marveling attracted him, also by long tradition, toward crystallizations of a subterranean sort. Scudéry, for example:

> There one finds a grotto of esteemed beauty,
> Where one sees a mix of light and dark,
> One hundred crystal rocks, jagged-tipped,
> Among stones encrusted with opals and rubies.[28]

Stendhal needed "a new word."[29] But he had to evoke a phenomenon that was not only well known to his readers, thus true and indubitable, but that would appear to them as both singular and mysterious as well. This is the case with the formation of crystals on objects immersed in a salty atmosphere surrounding. In Stendhal's time, the phenomenon was familiar, travel chronicles made frequent mention of it, and one comes across allusions to it more or less everywhere.

I have already pointed out how geometric crystalline shapes suited the Stendhalian preference for a logical explication, one with the rigor of a technical working drawing. But why did he choose *crystallization* rather than *sublimation*?

The two terms are close. In some passages from novels to which I will return, Victor Hugo employs terms that are derivative of *sublimation* to describe a more prolonged, more complete crystallization. It takes place in the depths: the earth is the matrix where this mysterious process of an ordinary substance—water or lead—transmuting into a crystalline gemstone is slowly worked out. Victor Hugo terms this transformation "sublimation." He conceives of it as a progression through the hierarchy of minerals, a hierarchy in which ice is inferior to crystal, which itself is subordinate to the diamond. Moreover, for Stendhal, the passion of love is likewise a movement of ascent: the imagination exalts the qualities of the loved being and raises it above its real state. This is indeed a matter of sublimation in the psychological sense of the term.

Of course, the physical process of sublimation, the direct, reversible transformation of a solid into a gas, permits collection of purified crystals. Since the substantive denotes not only the process of crystal formation but also the result of this process, an individual grouping of crystals bears the name "crystallization." It was thus nearly inevitable that Stendhal, with the presuppositions on which he relied, should choose this term.

He likewise sets for himself the question of the functioning of amorous crystallization, which he defines thus: "This phenomenon which I have allowed myself to call crystallization, arises from the promptings of Nature, which urges us to enjoy ourselves and drives the blood to our brains, from the feeling that our delight increases with the perfections of the beloved, and from the thought: 'She is mine.' " (6).

Thus there occurs in succession an elevation, an expansion, and an appropriation. The process begins with an impulse to satisfy desire; following this is a quantified aspiration to the greatest pleasure; finally, the notion of possession takes hold (which, incidentally, is close to the Freudian hypothesis of a condensation process, *Verdichtung* in German).

Crystallization "takes place in the depths." The image is one of a bough that becomes spangled with salt crystals in the very depths of the Salzburg salt mine. In the earth's heart, the vegetal becomes mineral or at least is covered with a mineral coating. But this concretion itself, in which, according to Stendhal, "the original bough is no longer recognizable" (6), retains from vegetal nature the able audacity to radiate out, with the ramified structure of branches that spread themselves out in space in segments or curves.

As a matter of fact, one also often finds this union (or this hybrid) of the vegetal and the mineral in the writings of chemical scientists, contemporaries of Stendhal, when describing or analyzing the phenomenon of crystallization. For example, Chaptal, following Petit and Rouelle, mentions the vegetation of salts: "that property they have of scaling the sides of vases that contain them in solution, . . . this is what is called *saline vegetation.*" I also borrow from Chaptal the description of the efflorescence of certain salts that "let [the water of crystallization] dissipate once they are exposed to air, such that soda, sodium sulfate, etc., and then these salts, lose their transparency, crumble into dust, and are called *effloresced* salts."[30]

The vegetal and the mineral have related structures that are branching and regular. Is it for this reason that Guillaume Davisson, a seventeenth-century chemist, extended the principle of crystallization beyond salts and minerals to the alveoli of hives and the leaves and petals of flowers?[31] Kepler, in his *Strena,* turned to the alveoli of bees, the symmetry of flowers, and the polyhedral appearance of pomegranate seeds to explain the six-angled strength of snowflakes.[32] In Stendhal's time, Father René-Just Haüy, the founder of mineralogy, repeated this comparison of the vegetal and the mineral, which allowed him to specify their differences precisely:

All [individuals of one and the same vegetal species] have flowers composed of parts equal in number and alike in shape and arrangement; the same relations hold between the respective positions of the leaves, in their rounded or angular contours, smooth or dentate. . . . If you have seen one, you have seen the entire species.

But the mineral is just an assemblage of similar molecules, joined by their resemblance; its growth takes place through the juxtaposition of new molecules, which apply to its surface, and its configuration, which depends solely on the arrangement of the molecules, can vary under the effect of differing conditions. Hence the multitude of different shapes, at once regular and sharply pronounced, that crystals of the same substance often assume.[33]

Stendhal's account takes its place within this tradition when, in his turn, he makes use of the mutual affinity of the vegetal and the mineral, which are brought together under wintry, clandestine conditions. The literal meaning is surely that one must rid the bough of its leaves and blooms so as to avoid their being withered by the surrounding brine. Salt crystallization then arrays the boughs in sparkling trimmings, a bit like the pearls strung on thin metal rods to make funeral wreaths. The symbolic meaning is that of the eternal return. It combines death and rebirth: winter is the time when vegetation is in its period of latency, when germination begins and prepares spring's explosion of green. Thus, in symbolic thinking—which is not so much prescientific as proscientific—there is a harmony between the vegetal bough and saline concretions, which in some form replace and perpetuate the vegetal ornaments, the lost leaves and blooms.

To return to common experience, the image Stendhal employs is further enriched by observation of trees in winter coated with ice. Their branches become encased in a gangue of ice, channeling and reemitting light from all directions. The forest is then simply a vast, soft glistening of enclosed light, diffused through all these delicate translucent sheaths.

One might object that besides the conjunction of the vegetal and mineral, the animal makes an appearance as well, since Stendhal also specifies that it is "the tiniest twigs, no bigger than a titmouse's claw" that are "spangled with a vast number of shimmering, glittering diamonds" (6). The sole function of this image is to indicate the extremely small size of the boughs. The titmouse is a very tiny bird, with spindly legs and claws; Stendhal the Anglicist likely knew its English name, which also evokes its small size. [A folk etymology of the word *tit* traces it to "(assumed) Middle English: any small object or creature."—Trans.] This passage thus indicates that these are minuscule branches covered in glittering crystals and, perhaps further, that in order to gain access to this hidden splendor, one must pay attention to and inspect them closely. Just like the lover who "discovers fresh perfections in the beloved at every turn of events" (6).

Stendhal's description of amorous crystallization shows another point of convergence with the description of the physical phenomenon of crystallization given by chemists who were his contemporaries: the requisite three conditions for crystallization are patience, rest, and the space within which to let it expand. Stendhal's text specifies these very points concisely: the time needed for crystallization around the branch is "two or three months" in "the depths of the disused workings" of the Salzburg mine. Stendhal further writes that "solitude and a period of quiescence [are] indispensable for the process of crystallization" (6).

The related notion of the very slow transmutation of metals within the earth, of lead into silver and then finally into gold, reigned at least until the Renaissance. It was founded on the frequent observation of silver in lead ores or gold in silver ores. This very lovely idea, of a gradual maturation of metals, termed "imperfect," into gold was strengthened, in fact, by the floral appearance of certain mineral concretions. It was anchored in the conviction that the "seeds" of metals came from emanations from the corresponding stars: Mercury for the metal of that name, Venus for copper, Mars for iron, the moon for silver, the sun for gold, Saturn for lead, and Jupiter for tin. With his notion of the crystallization of love, Stendhal rejoins this ancient myth of the hidden germination of metals, crystals, and precious stones. It is a concept that goes back a very long way: for example, rock crystal was taken by Seneca to be frozen water.[34]

I borrow from Stendhal's contemporary, Victor Hugo, these two expressions of the old myth. In *The Laughing Man*, Hugo writes: "Crystal is sublimated ice and . . . diamond is sublimated crystal; it is a known fact that ice becomes crystal in 1,000 years, and that crystal becomes diamond in 1,000 centuries."[35] In *The Hunchback of Notre Dame*, composed well before *The Laughing Man*, Victor Hugo already had the following affirmed by the deacon Charles Frollo: "Alchemy has its actual discoveries. Can you contest such results as these, for instance—ice, buried underground for one thousand years, is converted into rock crystal; lead is the progenitor of all metals (for gold is not a metal, gold is light); lead requires but four periods of two hundred years each to pass successively from the condition of lead to that of red arsenic, from red arsenic to tin, from tin to silver. Are these facts, or are they not?"[36]

For Stendhal, the crystallization of salt on a branch was emblematic of nascent love. In another version of such symbolic thinking, for Ramakrishna, a salt figurine dissolving in the ocean is itself emblematic of love coming to an end, in a return to the bosom of God.

RAMAKRISHNA'S EMBLEM.   Consider this parable from the Hindu mystic Ramakrishna (1836–1886):

> One day Rishi Krishna was walking along the seashore when one of his disciples approached Him and said to Him: "Lord, how can one reach God?"
>
> The Lord went down into the water and plunged him under. After a bit, He let him up, and, taking him by the arm, He asked him: "What did you feel?"
>
> The disciple replied: "I felt like the life was leaving me. My heart beat as to burst. I tried frantically to breathe."
>
> Then the Lord said to him: "You will see the Father when your thirst to see Him is as intense as was your need to breathe. God is like an ocean of sweetness; do you not want to dive deep into that ocean? If you dive into the Divine Ocean, you need fear neither danger nor death.
>
> "There was a doll of salt that wanted to take the measure of the ocean's depths. To this end, it took along a sounding line and a weight. It reached the water's edge and contemplated the powerful Ocean that stretched out before it. Until that moment, it still remained the same salt doll and retained its own individuality. But hardly had it set foot in the water when it became nothing but one with the Ocean. It was lost, all the salt particles it was made of were dissolved in the seawater. The salt it was made of came from the Ocean, and, just like that, it had returned to the Ocean, to join with it once more. . . . The salt doll dives into the ocean to take the measure of its depths, but as soon as the water touches it, it dissolves. So who will come tell us the depth of these ocean abysses?"[37]

This apologue by Ramakrishna, rich with meaning, has the implicit characteristics of an emblem. Recall that an emblem is characterized by a triple structure linking an image (*figura*), a heraldic device or a title (*inscriptio*), and a commentary, or *suscriptio*. Here, the *figura* is the salt doll dissolving in the ocean; the title could be "The Salt Doll"; and the explanatory commentary, the moral of the fable, is "do likewise, abandon your individuality, let yourself dissolve and return to God."[38] Thus Ramakrishna's text—which I hope I do not betray too much in analyzing it on the basis of the translation given above—is indebted to the general principles of emblematic analysis, as is an advertising poster, a cartoon, or a frame of a comic strip.[39]

Ramakrishna's emblem is figured by a homunculus. But contrary to the archetype—traditional in fantasy tales—of the golem, who is brought to life and then takes to living autonomously from its creator, this one is called to a sort of death, a disintegration. There is a double movement of disembodiment: the first when the flesh-and-blood person accepts, on Ramakrishna's urging, to be represented by a salt doll (a mineralization of the organic); the second when this miniature dissolves in the ocean and returns to the bosom of nature or God (a dematerialization of the mineral).

If one returns to the image of the salt doll, attached to a rope, sinking toward the bottom like a leaded sounding line, and dissolving as the seawater eats away at it, one can see that the doll is subject to two forces or constraints: gravitational force, acting here as a metaphor for the invincible attraction that the creature feels for God, and the dissolving power of water on salt, whose corresponding emblematic element is the moral injunction to let oneself enter the infinite goodness of God. This is the mystical theme of the dissolution of the personality, of the abjuration of individualism, the renunciation of the self.

If the emphasis in this text is placed instead on the liquid medium into which the salt doll dives (or into which it is plunged, for mystical writings are sometimes very prescriptive), it is an example of sacred water, of lustral water: we are witnesses to a redemptive baptism.

conclusion **ethics and politics**

The exploitation of salt amounted to an even greater exploitation of human beings. One sees the ready-made advantage to the sovereign, since this good lends itself so readily to taxation. But why did people allow themselves to be exploited to such an extent (for the uprisings, the rebellions against the gabelles, were only sudden, brief outbursts scattered over the long term of history)? This reference to the social contract, and to the distortion that salt brings to it, rendering it at once inegalitarian and unjust, raises the question of power: why seek it? why suffer it?

The answer bears on the biological, on the dominant–dominated relation one finds in a number of animal species. In human societies, salt supplied a foundation for lasting dominations.

In classical times, the extraction of salt was not only due to settled populations but was the occupation of slaves. A chance parenthetical remark of Cicero's reveals that this state of affairs was too familiar to his readers for him to have needed to explain it to them. And we are not entirely removed from this state of affairs: if petroleum can be called the salt of our times, are the peoples of the OPEC nations, from Venezuela to Gabon and Kuwait, enslaved peoples or peoples of slaves? To assert this would be simplistic, much too cut and dried. Nevertheless, it is not completely without ground.

Another aspect, commonplace but worth noting, of the commercial exchange of salt is that the tax constitutes the major part of the surplus value. On this point, one hesitates between two stances, two discourses, each one legitimate and grounded, namely, that of the political Left (surplus value is immoral, since it is collected at the expense of everyone, including the most disadvantaged) and that of the Right (surplus value—let's call it a return on investment—is moral since it alone permits innovation).

In this age of globalization and of an economism that prevails absolutely, this second view rules. Of course, it is no longer a question of salt but of extremely fluid capital accumulations resulting from the work of workers who are salaried, a word, let's recall in passing, derived from Rome and its legionnaires, whose pay was issued partly in cakes of salt. Thus the assumption of risk by innovative, small and medium-sized, high-tech businesses—that is, in one of the most dynamic sectors of the economy—takes place thanks to U.S. pension funds, rich to the tune of thousands of billions of dollars.

This last claim sums up the conventional account, an elitist and technocratic one, that presently holds sway. Does it take us very far from fourteenth- and fifteenth-century Venetian precapitalism, which invested in the building and fitting out of vessels sent to load up with exotic goods in the Orient and realized profits through the sale of salt to other northern Italian cities?

So, an economic perspective on salt in history obscures this merchandise, this foodstuff of prime necessity, behind the tax that has burdened it so frequently. One hardly sees the salt for the gabelle!

POPULARIZATION.   The role of salt in science, and the writings on popular science to which this role gave rise, calls for a similar objection. Once again, salt, despite its corporeality, its density, its refractory quality (with high melting and vaporizing temperatures), in short, all its rich, positive nature, appears like a compound at once fake—nearly fictional—and ungraspable.

I will point out two aspects of such difficulty in conceptualization. First, our empirical knowledge presents salt to us as a white solid, as one of the most common manifestations of whiteness. This de facto condition is not required in theory, since purified salt, far from reflecting and diffusing light (whence that white appearance), is actually transparent. And which of these should physics and chemistry treat: salt in its Platonic archetype as a transparent entity or salt in its ordinary and familiar form as a white solid? Must science study an impure but existing solid or an ideal and nearly unrealizable entity?

This same question arises with regard to the second aspect: in ideal form, sodium chloride is a rigid solid, its atoms arranged in a perfectly regular, crystalline network. In actuality, their positioning is imperfect: a salt crystal is a virtual fluid, ready to flow down fracture lines and surfaces (a nice image and metaphor for the current state of our societies?). What, then, is the object of study? Is it superb, immutable Platonic structure, or is it the proliferation of varying Aristotelian kinds, which among other things are subject to this phenomenon of flow? It is somewhat annoying, and troubling, as well, that science should enclose itself in such ivory—and salt—towers. Can there be science only in and through abstraction, purification, and Puritanism? Is all possibility of knowledge about the objects of daily life ruled out in principle? And what about instruction? Does this prohibit us from helping our children to discover a given phenomenon with the aid of bits of string and matches, at the cost of later introducing them to the corresponding mathematical expressions, which necessarily concern more evanescent ideal entities?

To conclude this discussion of the distance between science and common sense (a distance that is necessary), I sought in the more scientific parts of the book—for example, in the section on the invention of spectroscopy—to bring out the unexpected and paradoxical moments of science, in that particular case, moving in one step from the ordinary gas burner to the chemical composition of the sun.

If I have placed the emphasis on salt, as a sort of revolving central point, a recurring object of scientific inquiry and subject of experiment, this is also because it provides an ideal topic for conveying science to nonscientists, for carrying out what our language commonly calls "the popularization of science."

Is this how one should popularize? And what does one speak of when one popularizes? My position in this book was to try to show that knowledge, like the nature that it describes and helps us to understand, if not explain, is all of a piece. Not only is it possible to proceed in this way, we also have a moral obligation to be generalists; we must move away from the specialization of our training toward a natural philosophy, the one that our epistemic period seeks with such difficulty to express.

THE REPRESENTATION OF HISTORY.   My final point for reflection is the following: What does one gain by tossing together memories, by recalling one or more stories, by presuming to write history? Doesn't it amount to drawing water in a basket, to writing in the sand? In fact, there exists an infinity of historical accounts. Relying on the same facts, these chronicles place them in different order and thus impart to them different meanings. History and science each have a bone to pick with relativism, an ideology all the more pernicious for occasioning the slide from "what's the use?" to "no one can prohibit me from saying anything whatsoever, since every point of view contains its own justification." In this regard, the Sokal affair proved a revelation to the general public, pointing out some often-irresponsible social studies of science. In blunt terms, will history of science die because of elitism or because of incompetence and lack of relevance? For the problem remains how to write history of science in a manner both well researched (which presupposes that the author has acquired the necessary technical competence) and readable (which demands jettisoning more technical stuff).

My answer is the work, is in and through the work itself. I will explain this by means of an image: Watercolorists frequently paint landscapes. They thus must produce depictions of skies, something very difficult: How is it possible to transfer to paper at once the gradation of the nuances of blue, the cottony diffusion of the clouds, the depth of the azure, and the vibration of the light that fills this whole vaporous, clear ether? One trick of the trade is first of all to drench the strokes of color in plenty of water; next, the artist sprinkles salt on his sky. And after the grains of salt have absorbed the nearby water— the excess water to be more precise—he removes the salt from the paper. The effect of the process seems like a kind of magic: a sky is revealed (what luck, it appears when summoned!). I see in this a metaphor for writing, when it is a question of approaching a truth that is a priori unreachable.

So, then, as a fitting end to this book, I will take the liberty of presenting the reader with the little painting that follows.

afterword **the union of earth and sea**

It is the wedding of earth and sea. They have become indistinguishable. Banks of sand are covered in a thin sheet of water. Rivulets separating them etch and grow with the currents and tides. An entire watery writing that one can examine from the minutest level to as far as the eye can see is thus remade, secretly and at length rather than plainly and at once. Arms of the sea become hands of the sea, fingers of the sea, hair of the sea. The earth, also tousled after its long congress with the tide, water streaming from every pore on every side, displays itself immodestly in the sun. Its body, still salty from their common passion, has become dotted with tiny whitish remnants in the carelessness of its transports: leavings of dried seaweed, shells bleached in the sun, a bit of foam here and there, come from we no longer know where.

notes

[Unless otherwise indicated, all translations are mine.—Trans.]

I. SALT-CURED FOODS

1. A plant example is that of weeds, which can be eliminated from between the stones or bricks of a terrace by sprinkling them with salt, or sodium, or potassium chlorate (the last is toxic, a powerful oxidizing agent).
2. Hence the English proverb "to rub salt into a wound," which has come to mean metaphorically "to lack delicacy, to offend." The correlative saying in Portuguese is "pôr o sal na moleira," literally, "to put salt on the molar," which means "to annoy someone."
3. Not only proteins but fats as well, as in the case not only of cheeses but also of salted butter.
4. Still fundamental to the economy, daily life, and culinary specialties of Portugal today.

5. M. A. Alexander and W. C. Stringer, *Country Curing Hams*, Agricultural publication G02526 (Columbia: University of Missouri–Columbia, 1993).

    Pig farming and the salt curing of pork usually take place in different locations: Ardennes hams (Belgium) chiefly come from pigs raised in Flanders; similarly, Bayonne hams, from the Basque region of the Adour River basin, sold in Bayonne, a town that has long prospered from commerce, are salted with salt from Salies-de-Béarn. On the occasion of the visit to Bayonne of the constable of Montmorency—who went there in 1529 to collect the ransom demanded by Emperor Charles V upon the liberation of François I—the city presented him with gifts, among which were salted hams; this is the earliest mention of the hams in the Bayonne municipal archives. Another anecdote relates how Catherine de Médici offered an adolescent Charles IX a tour of his kingdom, with the political aim of consolidating its borders. On a visit to the Bayonne saltworks, she is said to have pronounced: "Behold your gold mines." See G. Dunoyer de Segonzac, *Les chemins du sel* (Paris: Gallimard, Découvertes, 1991), 80. See also Louis Laborde-Balen, *Le livre d'or du jambon de Bayonne* (Pau: Cerpic, 1991).

6. *Grote Winkler Prins encyclopedie* (Amsterdam: Elsevier, 1993), s.v. "*haringkaken.*"

7. Some sources trace its origin to Abö, Finland, in the thirteenth century; see *De Katholieke Encyclopaedie* (Amsterdam, 1950), s.v. "Beukelszoon, Willem," 854.

8. *Grote Winkler Prins encyclopedie* (Amsterdam: Elsevier, 1993), s.v. "Beukelsz, Willem," 164; Abraham van der Aa, *Biographisch Woordenboek der Nederlanden* (Haarlem, 1852; reprint, Amsterdam: B. M. Israël, 1969), s.v. "Beukelsz, Willem"; *Grote Nederlandse Larousse Encyclopédie* (s'-Gravenhage, 1971), s.v. "Beukelsz, Willem," 627; *De Grote Oosthoek* (Utrecht, 1965), s.v. "Beukels, Willem," 240; *Standaard Encyclopedie* (Utrecht, 1965), s.v. "Beukels, Willem"; *De Katholieke Encyclopaedie*, s.v. "Beukelszoon, Willem," 854.

9. This is what I gleaned from the aforementioned encyclopedias; the last one seems to be the least serious.

10. The innovation spread rapidly. Beginning in 1409, Burgundian regulations report the following:

    Jehan, duke of Burgundy, count of Flanders, etc. . . . To our bailiffs in Bruges, l'Eau à l'Ecluse, Nieuport, Ostende, Biervliet, and all the other bailiffs, dispensers of justice, and officials in our said county and land of Flanders or to their lieutenants, greetings. We have received the humble petition of the sailors and fishermen of our said lands and county of Flanders, claiming that, in order to ensure their life and sustenance and that of their wives and children, it is their habit to fish beyond the Waal and each year from the feast of Saint Bartholomew [August 24] to the feast of Saint Matthew [September 21] to take fresh herrings and salt them in casks to make salt-cured herring, which is fitting and good; but they neither can nor have been able to sell them in our land of Flanders . . . save in our town of Biervliet; for this reason, they received favour and permission from us, through our letters patent, issued in our city of Bruges the seventh day of September in the year 1409, to bring the herring here for sale or to have them sold here, in exchange for a payment to us of one noble per load of herring. (Cited by Michel Mollat, *Genèse médiévale de la France moderne, XIVe–XVe siècle* [Paris: Arthaud–Le Seuil, Points, 1977], 110.)

11. Tosh Lubek and Gary Stein, "Historical Overview: The Essential Guide to Aberdeen," *The Essential Guide to Aberdeen*, http://www.webit.win-termute.co.uk/webit/index.htm, 1996.

12. "The Dutch also came to the Shetland Isles in June . . . for their herring fishing. . . . This trade is highly profitable for those who devote themselves to it; it is often said that fishing is responsible for the level of wealth and eminence that the Netherlands has attained: thus, certain historians describe fishing as the gold mine of Holland" (J. Brand, *A Brief Description of Orkney, Zetland etc.* [Edinburgh, 1701]). Fleets comprising as many as twenty-five hundred Dutch fishing boats, or twenty thousand men, assembled at the Shetlands in mid-June, from there to follow the herring migration into the Channel. Michelet devoted a celebrated page to this in *La Mer*, which I have shown originated in an entry from Diderot and d'Alembert's *L'Encyclopédie*; see P. Laszlo, "La fabrique du texte chez Michelet," *Poétique* 70 (1987): 219–30.

13. Alexandre Dumas, *Voyage en Russie* (Paris: Hermann, 1960), 667.

14. August von Haxthausen, *Studies on the Interior of Russia* (n.p., 1847).

15. François Cardarelli, *Scientific Unit Conversion* (London: Springer, 1996), 66.

16. M. Jouvé, "Pasteurisation par champ électrique pulsé" (Auch: CRITT [Centre régional d'innovation et de transfert de technologie], 1997).

17. R. C. Whiting and P. Fratamico, "Reduction of Spoilage and Pathogenic Microorganisms in Foods with a Parasitic Bacterium, *Bdellovibrio bacteriovirus*" (Wyndmoor, Pa.: Easter Regional Research Center, ARS, USDA, 1997). For example, adding *Bdellovibrio bacteriovirus* to meat products (such as fowl) is advised. This bacterium is an inevitable parasite of gram-negative bacteria such as *Shigella, Yersinia, Escherichia coli*, and *Vibrio*, as well as of various salmonellae. *Bdellovibrio* is aerobic, and it prolif-

erates at a neutral pH and ambient temperature. It should be harmless to human beings.

18. This simply means "salted beef." In English, the three words *grain, kernal,* and *corn* are derived from the same Indo-European root **grH-*, which is found in French in *grain* or *gravier.* Also, in Old English, *corn* means salt in grain form. See Louis Heller, Alexander Humez, and Malcah Dror, *The Private Lives of English Words* (London: Routledge and Kegan Paul, 1984), 52.

19. In certain countries (Switzerland), one finds powders made of sodium glutamate for sale for the enhancement of flavors in foods.

20. There must have been many varieties of it, just as today there exist many versions of the cold soup called gazpacho. On this last point, see, for example, Alice B. Toklas, *The Alice B. Toklas Cookbook* (New York: Harper, 1954).

21. Pliny, *Natural History*, book 31, 87 and 93–94; Bertrand Guégan, ed., *Apicius: Les Dix Livres de cuisine*, trans. René Bonne (Paris: Bonnell, 1933). Horace also mentions *garum* in *Satires*, book 2, satire 8.

22. Joseph B. Dunphy, *Ancient Roman Fish Sauce* rec.food.historic, June 13, 1997.

23. The Niçoise dish *pissala* (whence is derived *pissaladière*) is made with whole small fish (sardines, anchovies, fish for frying, etc) placed for salting with fish offal, small shellfish in an embryonic state, and fish roe in a salting tub with salt and fines herbes. See Maguelonne Toussaint-Samat, *A History of Food* (Cambridge, Mass.: Blackwell Reference, 1993), 374. Today, *boutargue* (or *poutargue*) is made of fish roe preserved in salt, "a dish made of fish roe: dried, salted, or smoked gray mullet, mullet, or bass. This is the Provençal caviar that the Venetians called *botaga*. Already known to the Cretans in the time of Minos, we believe it was imported into Provence by the Phoenicians when they founded Marseilles in the sixth century before Jesus Christ" (Lucette Rey-Billeton, *Les bonnes recettes du soleil: Richesses des terroirs* [Avignon: Aubanel, 1984], 1:44). "The Sardinians, who have immense cauldrons of sea water at the very south-east end of their island, could doubtless link these to their celebrated *botargo*, fish roe (mullet) that are pressed and salted by the Mediterranean. (This delicacy, well-known in Jacobean England, was served at the coronation feast of James II in 1865.)" (Elizabeth David, *Spices, Salt and Aromatics in the English Kitchen* [Harmondsworth, Middlesex: Penguin, 1970], 52).

24. Elizabeth Duran Gessner, *Ancient Roman fish sauce* rec.food.historic, April 30, 1997.

25. This dinner was given in the town of Bonnieux on Saturday, July 19, 1997, for the Journées romaines (Roman days); it was held at the Pont Julien as a part of its bimillennial celebration and was organized by the villages of the Apt district.

26. Pierre Belon, a seventeenth-century French explorer and naturalist, described a *garum* he saw prepared in Constantinople, particularly in the city's Pera quarter. See Andrew Dalby, *Siren Feasts: A History of Food and Gastronomy in Greece* (London: Routledge, 1996), 200.

27. Aimé Vayssier, *Dictionnaire patois-français du département de l'Aveyron* (1879; reprint, Marseilles: Lafitte Reprints, 1979).

28. *Le nouveau cuisinier royal et bourgeois* (Paris: Claude Prudhomme, 1721), 2:327.

29. Jacques Boisgontier, *Dictionnaire du français régional du Midi toulousain et pyrénnéen* (Paris: Bonneton, 1992).

30. Hervé This, *Les secrets de la casserole* (Paris: Belin, 1993), 118–33.

31. Salty appetizers are served with apéritifs. They fall into the category of the fruits of a harvest, or of salt-cured foods, or into a third category of equivalents of the salt-bread pair: almonds, hazelnuts, pistachio nuts, cashew nuts, Brazil nuts, and macadamia nuts; green and black olives; raw vegetables served with a sprinkling of salt or a dipping sauce; winkles, shrimp, oysters, and other seafood; small pieces of cheese; cold cooked meats; crackers of the "Triangolini" or "Tuc" sort, pretzels, and so on. Sometimes, the alcoholic beverage served to welcome guests is itself directly associated with salt: think of tequila or a margarita, drunk in a glass whose rim has first been coated with salt.

32. For example, pork chops, pig's tails, ham, veal shank, beef sirloin strip, beef shoulder, bottom round of beef, duck filet, chicken breast, turkey legs, sea-bream filets, mutton chops, leg of lamb, octopus, saveloy, Toulouse sausage, hare, river fish, tuna, turbot, cod, snails . . .

33. Cardoons, asparagus, salsify, onions, peas, peppers, green beans, beets, potatoes, white beans, red beans, black beans, lentils, tomatoes, cucumbers, eggplants, squash, wild mushrooms, broad beans, artichokes, chestnuts, oranges, apples, pears . . .

34. Broth, béchamel, gribiche, hollandaise, *à l'armoricaine*, white wine, red wine, Madeira, Port, rouille, mayonnaise, aïoli, and pauvre homme!

35. To limit ourselves to the most current ones: curry, paprika, harissa, nutmeg, cloves, cinnamon, cumin, coriander, saffron, cayenne, aniseed, allspice . . .

36. Thyme, savory, parsley, sage, rosemary, tarragon, dill . . .

37. Boiled, fried, stuffed, roasted, braised . . .

38. In dress also, Father Christmas's red robes recall the episcopal robes of the saint.

39. This is the date of death of Saint Nicholas, whose legend is linked to a historical figure: the bishop of Myra, in Asia Minor, was imprisoned with a good many other Christians by Diocletian in 303 ACE before being liberated by a decree from Constantine; Nicholas participated in the Nicene Council in 325 and died December 6, approximately 343.

See Charles W. Jones, *Saint Nicholas of Myra, Bari, and Manhattan: Biography of a Legend* (Chicago: University of Chicago Press, 1978).

40. The second couplet of the song *Saint Nicolas patron des écoliers* (Saint Nicholas, patron saint of school children) goes: "They had hardly entered / When the butcher killed them . . . / Cut them into little pieces / Put them in the salting tub like swine!" (Philippe Duley, *Saint Nicolas: Prestiges de L'Est* [Meurthe et Moselle: Editions de l'Est, 1990]).

41. Here I follow the highly interesting work of Colette Méchin, *Saint Nicolas: Fêtes et traditions populaires d'aujourd'hui*, Espace des hommes (Paris: Berger-Levrault, 1978).

42. "The whips of Lower Silesia and the Flemish sticks wielded by these animal maskers have . . . a fertilizing nature. They are very similar to the lashes made of the leather from a sacrificial ram that were used by the wolf-men of ancient Rome to strike women to make them fertile" (ibid., 96).

43. "To discern, beneath their burlesque conduct, an important function: that of permitting the dead, guarantors of wealth and fecundity, to return for a time to earth" (ibid., p. 94).

44. An explanation, which I haven't found in the literature but which would require philological analysis to support it, is the assimilation of the place name *Myra* (Saint Nicholas was its bishop) with the name for *brine* in Latin, *muira*. It could be compounded by another lexical confusion, that between *Myra* and *myrrha*, or myrrh. Moreover, myrrh (an aromatic plant, one of the gifts of the magi to the Christ child) served, as did salt, to embalm the dead in antiquity, particularly in Egypt. Thus a pun could be at the origin of the thematic of death and the salt curing of the children in the legend of Saint Nicholas.

45. "This flesh has the quite astonishing ability to link Christmas with the Carnival cycle" (Méchin, *Saint Nicolas*, 46).

46. Although salt production in Dombasle-Varangéville dates only to 1853, the Lorraine region already had many ancient saltworks, which its place names still reflect. It also had a very ancient salt route in use since antiquity that linked Toulois and le Vermois to Saulnois. See ibid., 135–39.

2. NOMADS

1. Herodotus, *The Histories*, book 4 (Harmondsworth: Penguin, 1959), 303–4. Translation adapted.

2. For evaporites, salt deposits formed by evaporation in arid climates, see Kjell T. Svindland, *Help, What Are Evaporites?* sci.geo.geology, May 24, 1997.

3. Jeremy Swift, *Le Sahara* (Amsterdam: Time-Life, 1975); see in particular the photograph on p. 127.

4. Bruno Verlet, *Le Sahara*, Que sais-je? (Paris: PUF, 1974). But this point, a commonplace in writings on the Sahara, is strongly contested by Odette Bernezat, *Hommes des montagnes du Hoggar* (Grenoble: Quatre Seigneurs, 1979), 362–63.

5. Hilde Gauthier-Pilters and Anne Innis Dagg, *The Camel: Its Evolution, Ecology, Behavior, and Relationship to Man* (Chicago: University of Chicago Press, 1981), 122.

6. Patricia Crone, *Meccan Trade and the Rise of Islam* (Princeton, N.J.: Princeton University Press, 1987).

7. Gauthier-Pilters and Dagg, *The Camel*, 118–22, 159–62, ch. 5.

8. Desmond Bagley, *Flyaway* (Glasgow: Fontana-Collins, 1979), 158.

9. *Pictures for Crusoe*, trans. Louise Varèse, in Saint-John Perse, *Collected Poems: Complete Edition with Translations by W. H. Auden, Hugh Chisholm, Denis Devlin, T. S. Eliot, Robert Fitzgerald, Wallace Fowlie, Richard Howard, Louise Varèse* (Princeton, N.J.: Princeton University Press, 1983), 63.

10. *Pour fêter une enfance*, trans. Louise Varèse, in Perse, *Collected Poems*, 23.

11. *Anabase*, trans. T. S. Eliot, in Perse, *Collected Poems*, 105.

12. *Exil*, trans. Denis Devlin, in Perse, *Collected Poems*, 157.

13. *Anabase*, trans. T. S. Eliot, in Perse, *Collected Poems*, 105.

14. The English translation of *Terre des hommes* is *Wind, Sand and Stars*, trans. Lewis Galantière (New York: Harcourt Brace Jovanovich, 1992). The English translation of *Citadelle* is *The Wisdom of the Sands*, trans. Stuart Gilbert (New York: Harcourt Brace, 1950).

15. *Anabase*, trans. T. S. Eliot, in Perse, *Collected Poems*, 105.

16. *Exil*, trans. Denis Devlin, in Perse, *Collected Poems*, 149.

17. *Anabase*, trans. T. S. Eliot, in Perse, *Collected Poems*, 103.

18. Jean-Pierre Richard, *Onze études sur la poésie moderne* (Paris: Seuil, 1964), 31–66.

19. Anonymous, *A Hand-book for Travellers in Switzerland and the Alps of Savoy and Piedmont* (London: John Murray, 1838; reprint, Leicester: Leicester University Press, 1970), xix.

20. H. Affre, *Dictionnaire des institutions, mœurs et coutumes du Rouergue* (1903; reprint, Marseilles: Lafitte Reprints, 1974). Each summer, Jean-Yves Bonnet, of F-12120 Salmiech, leads a reenactment on horseback of the trip along the Languedoc salt route from Valras to Rodez by way of Bouloc. In 1291 Philippe le Bel acquired the saltworks of Peccaïs along with other Languedoc saltworks, a change in ownership that is not incidental to the 1317 establishment of the salt gabelle in his kingdom, a tax reimposed in 1341. See Jacques Le Goff, "Le sel dans les relations internationales au Moyen Age et à l'époque moderne," in *Le rôle du sel dans*

l'histoire, ed. Michel Mollat, Publications de la faculté des lettres et sciences humaines de Paris-Sorbonne, Recherches, vol. 37 (Paris: PUF, 1968), 235–45.

21. E. Baratier et al., *Atlas historique* (Paris: Armand Colin, 1969), 44.

22. For this chronicle of the Nice–Turin Real Strada, see Jean-Loup Fontana and Michel Foussard, *Real Strada: La route royale de Nice à Turin* (Nice: CAUE Alpes-Maritimes, 1993).

23. Baratier et al., *Atlas historique*, 44.

24. Karl Baedeker, *Le Midi de la France depuis l'Auvergne et y compris les Alpes*, 4th ed. (Leipzig: Baedeker, 1892), 434. Judging by the meal prices and hotel rates listed in the guide, one must multiply prices by a factor of approximately twenty.

25. Baratier et al., *Atlas historique*.

26. Baedeker, *Le Midi*, 414.

27. One example is the order dated May 12, 1734, issued by Charles Alphonse Dalmazzone, the gabelle administrator and registrar-general of the county of Nice, concerning the requisitioning of pack animals for salt transport on the new Saint-Martin road (note on the number of animals and the required course to take for requisitions made in Villeneuve d'Entraunes, Alpes-Maritimes Departmental Archives, 02AFF 0201).

28. Cited in *Les Chansons du Carrateyon*, published in Aix.

29. As additional evidence, I include the Irish proverb "Ni chailleann an salann a ghoirteamas, ach cailleann na daoine an carthannas" (Salt does not lose its saltiness, but people lose their sense of charity). See T. A. O. Maille, *Sean-Fhocla Chonnacht*, no. 887 (Dublin, 1948), 1:135.

30. It is the pronunciation of an ell in the weak or implosive position at the end of a syllable. This phonetic phenomenon took place in the French language during the seventh to tenth centuries but is no longer occurring. It is still under way in Portuguese (more so in Brazil than in Portugal; think of the Brazilians' pronunciation of the name of their country) or in Polish, in words like Łodz [wudz^j]. An X-ray cinematographic image, made by Denise Jacquemin (who supplied me with this information) at the Grenoble University Hospital Center, compares the tongue movements during vocalizing of the syllable -*val* in French and in Portuguese; the transitory sound *o*—audible at the very end of Georges Marchais's declaration, "C'est un scandale!"—requires two sorts of tongue tensions: apical and velar. If the velar tension predominates, the contact at the back of the mouth disappears, producing the velarization specific to the *w* and *u* sounds.

31. The various examples are drawn from *Le Petit Larousse illustré 1995* (Paris: Larousse, 1994). For the verb *saumater*, see Pierre Lemonnier, *Paludiers de Guérande: Production du sel et histoire économique*, Mémoires de l'Institut

d'ethnologie XXII (Paris: Institut d'ethnologie, Musée de l'Homme, 1984), 244. On the subject of the town of Lons-le-Saunier, note that the saltworks there were no longer in operation after 1317–1320 because they were not profitable. See André Hammerer, *Sur les chemins du sel: activité commerciale des sauneries de Salins du XIVᵉ au XVIIIᵉ siècle* (Besançon: Cêtre, 1984).

32. This and subsequent terms discussed in this section are from Louis Alibert, *Dictionnaire occitan-français selon les parlers languedociens* (Toulouse: Institut d'études occitanes, 1993); and Jacques Boisgontier, *Dictionnaire du français régional du Midi toulousain et pyrénnéen* (Paris: Bonneton, 1992).

33. This and subsequent terms discussed in this section are from Lemonnier, *Paludiers*, 241–45.

34. Jean-Claude and Jacqueline Hocquet, "Le vocabulaire des techniques du marais salant dans l'Adriatique au Moyen Age," in *Mélanges de l'Ecole française de Rome*, 86, no. 2 (1974): 527–52.

35. This and subsequent terms discussed in this section are from the article "Saline," in *Dictionnaire raisonné des sciences, des arts, et des métiers à partir de 1751*, ed. Denis Diderot and Jean le Rond d'Alembert, 557–64 (confirmed on certain points by René Locatelli, "Du nouveau sur les salines comtoises au Moyen Age," *Société d'émulation du Jura: Travaux 1989* [Lons-le-Saunier, 1990], 153–68); and Claude-Isabelle Brelot and René Locatelli, *Un millénaire d'exploitation du sel en Franche-Comté: Contribution à l'archéologie industrielle des salines de Salins* (Besançon: CDRP, 1981).

36. *Brine, bryne*: according to the *Oxford English Dictionary*, a confirmed usage from the year 1000 in Old English in Ælfric's *Gloss*, in Wr.-Wülcker 128.

3. HARVESTING

1. Salt domes are sometimes found in association with oil deposits. The first discovery of this frequent salt-oil association dates from 1901, off the coast of Texas. In this area of the Gulf of Mexico, oil exploration and production has taken advantage of the association: in 1965 nearly 200 of the 329 domes investigated yielded a positive result.

2. Experimentation on sodium chloride monocrystals shows that creep, that is, this plastic deformation, is due on the microscopic scale to dislocation slippage, that is, to atomic lacunae in the crystalline network, along the reticular planes of the crystal, as on a moving walkway. These laboratory studies, and the crystalline microstructures that they have identified, systematized, and explained, allow us to reconstruct the geophysical history of salt domes. See J-P. Poirier, *Philosophical Magazine* 26

(1972): 701–12, 713–25; J-P. Poirier and G. Martin, *Philosophical Magazine* 26 (1972); M. Guillope and J-P. Poirier, *Journal of Geophysical Research* 84 (1979): 5557–67.

3. The word *diapir*, from the Greek verb *diapirein*, "to pierce," was introduced by L. Mrazec in 1915.

4. For two excellent popular science articles, see Christopher J. Talbot and Martin P. A. Jackson, "Salt Tectonics," *Scientific American* 255 (1987): 70–79; and J. D. Martinez, "Salt Domes," *American Scientist* 79 (1991): 420–31.

5. Jules Conan, *Trésor scientifique des écoles primaires* (Paris: Delagrave, 1880), 130.

6. Montesquieu, *Considérations sur les richesses de l'Espagne*, ed. Charles Vellay (Paris: Jacques Bernard, 1929), 89.

7. François Jullien, *Eloge de la fadeur* (Arles: Philippe Picquier, 1991), 102.

8. Alain Colas, *Le sel*, Que sais-je?, no. 339, 2d ed. (Paris: PUF, 1993), 53.

9. Dampier Salt Ltd., *Rainfall and Evaporation*, http://www.dampiersalt. com.au/tnpn002785/prod/dsl/dslweb.nsf/LinkDocuments/Products, 1996.

10. P. Lemonnier, *Paludiers de guérande: Production du sel et histoire économique* (Paris: Institut d'ethnologie, Musée de l'Homme, 1984). The name Guérande means "white region" in the Breton language, from *gwen*, or "white," and *ran*, or "region." In the middle of the last century, two thousand saltworkers practiced their craft in the Guérande. Today, they number no more than two hundred, but they have pursued an up-market type of production and quality ("Red" label, "Nature et Progrès" brand) that guarantee them a sale price on the order of 1,300 francs per ton. See Eric Fottorino, "La liberté retrouvée des paludiers de Guérande," *Le Monde*, September 2, 1997, 17.

11. Honoré de Balzac, *Béatrix*, trans. James Waring (New York: Macmillan, 1901), 2.

12. Anonymous, "Desalination of Water and Sea-Water," pamphlet available from World-Water, 10201 N. Concord Drive, Mequon, WI 53097, 1997.

4. ABUSE OF POWER

1. Mary Douglas and Baron Isherwood, *The World of Goods: Towards Anthropology of Consumption* (New York: Routledge, 1996), 61.

2. The practice of drying fish on the shore can be traced to the start of the seventeenth century; see Sebastian Junger, *The Perfect Storm: The Story of Man Against the Sea* (New York: Norton, 1997), 41–42.

3. In the nineteenth century, American sailors still sang *The Salt Horse Song*:

"Oh horse, oh horse, what brought you here?
You carted stone for many a year

With kicks and cuffs and ill abuse
Now salted down for sailors' use

There you find him in damn great junks
Between the mainmast and the pumps.
You pick him up with great surprise
And throw him down and damn his eyes;

And gnaw the meat from off the bones
And throw the rest to Davy Jones.
Now if you don't believe this story's true
Look in the barr'l and you'll find his shoe."

From *Ballads Migrant in New England,* by the collectors Helen Hartness Flanders and Marguerite Olney, with an introduction by Robert Frost, New York: Farrar, Straus and Young, 1953, 226.

4. Joel Kotkin, "Cities Can Shine if They Take up Tools and Build on Renaissance Lessons," *International Herald Tribune,* August 20, 1997, 9.
5. Florence's strategic reply was precisely its acquisition in the fifteenth century of a maritime force and an independent means of supplying it with salt. See Jacques Le Goff, "Le sel dans les relations internationales au Moyen Age et à l'époque moderne," in *Le rôle du sel dans l'histoire,* ed. Michel Mollat, Publications de la faculté des lettres et sciences humaines de Paris-Sorbonne, Recherches, vol. 37 (Paris: PUF, 1968), 243.
6. Strabo, *The Geography of Strabo,* Vol. III, The Loeb Classical Library, trans. Horace Leonard Jones (New York: G. P. Putnam's Sons, 1917), 271.
7. Aristotle, *The Rhetoric and the Poetics of Aristotle,* introduction by Edward P. J. Corbett (New York: Random House, 1984), 150.
8. Peter Lauritzen, *Venice: A Thousand Years of Culture and Civilization* (London: Weidenfeld and Nicolson, 1978); Mary McCarthy, *The Stones of Florence: Venice Observed* (Harmondsworth: Penguin, 1972).
9. Jean-Claude Hocquet, *Le sel et la fortune de Venise,* vol. 1, *Production et monopole,* vol. 2, *Voiliers et commerce en Méditerranée, 1200–1650* (Lille: Presses universitaires de Lille, 1978, 1979); Michel Mollat, in *Venezia del Mille* (Venice, 1965).
10. Lauritzen, *Venice.*

11. Jean-Claude Hocquet, *Atti dell'Instituto Veneto di Scienze, Lettere ed Arti* 128 (1979): 525–74.

12. Giovanni Boccaccio, *The Decameron*, second ed., trans. G. H. McWilliam (New York: Penguin Classics, 1995), 632–33.

13. Bernard Meunier, "Le rôle du sel dans la civilisation," *Economie-Géographie* 306 (1993); Michel Mollat, ed., *Le rôle du sel dans l'histoire*, Publications de la faculté des lettres et sciences humaines de Paris-Sorbonne, Recherches, vol. 37 (Paris: PUF, 1968).

14. At the beginning of the thirteenth century, Venice seized the saltworks of Cervia that formerly had been fought over by Ravenna and Ferrare; at the beginning of the fourteenth century, Venice became the sole supplier of Mantua, a town that in the thirteenth century had attempted to liberate itself from Venetian salt. See Le Goff, "Le sel," 242–43.

15. The rivalry between these two cities took place especially in Lombardy. In the second half of the twelfth century, Genoese expansionism enjoyed great success, with the domination of the Hyères saltworks and the expulsion of the Pisans from the Sardinian salt production centers. See Le Goff, "Le sel."

16. Like northern Italy, the Netherlands experienced a strong demand for salt because of its population boom; see Le Goff, "Le sel," 240–41. But its supremacy in shipping Atlantic salt to northern Europe and the Baltic was only established in the sixteenth century and did not last beyond the end of the seventeenth century, when it was assaulted by English, Scottish, and Scandinavian shipping. See also Pierre Jeannin, "Le marché du sel marin dans l'Europe du Nord du XIVe au XVIIIe siècle," in *Le rôle du sel*, ed. Mollat, 73–93.

17. Meunier, "Le rôle du sel"; Hocquet, *Le sel*, vol. 2.

18. For many years, the Bay of Bourgneuf had the greatest production output of the Atlantic saltworks. Judged to be of lesser quality, its salt was 30 to 50 percent less costly than the competing salts from the English wiches, in particular from Cheshire, or than salt from Lüneburg. Then salt from Brouage became a rival. Finally, around 1770, Portuguese salt from Setúbal and Aveiro gained ascendancy. See Mollat, *Le rôle du sel*, 14, and Jeannin, "Le marché du sel marin," 75.

19. Jeannin, "Le marché du sel marin," 74.

20. See René Locatelli, "Du nouveau sur les salines comtoises au Moyen Age," *Société d'émulation du Jura: Travaux 1989* (Lons-le-Saunier, 1990), 153–68.

21. Lauritzen, *Venice*.

22. Meunier, "Le rôle du sel."

23. McCarthy, *Stones of Florence*, 197.

24. "Salt Trade and Industry," *Encyclopedia Judaica*, (Jerusalem, n.d.), 712–13.

25. A. R. Mitchell, "The European Fisheries," in E. E. Rich and C. H. Wilson, eds., *The Cambridge Economic History of Europe* (Cambridge: Cambridge University Press, 1977), 5:80–181.

26. H. G. Koenigsberger, "Western Europe and the Power of Spain," in *The Counter-Reformation and Price Revolution, 1559–1610*, vol. 3 of *The New Cambridge Modern History*, ed. R. B. Wernham (Cambridge: Cambridge University Press, 1968), 269.

27. J. H. Parry, "Transport and Trade Routes," in *The Economy of Expanding Europe in the XVIe and XVIIe Centuries*, vol. 4 of *The Cambridge Economic History of Europe*, ed. E. E. Rich and C. H. Wilson (Cambridge: Cambridge University Press, 1967), 184–85, 203–4. The island closest to the African continent in the Cap Verde archipelago is still named Ilha do Sal (Salt Island) because ships (in particular slaveships) stopped there to pick up salt. Ilha do Sal, known to Arab navigators, was "discovered" in 1460 by Diego Gomes and Antonio da Noli of Portugal. Its salt lake is located inside the crater of an extinct volcano in Pedra de Lume. Salt production there ceased in the first half of the twentieth century.

28. See the fascinating work of Emmanuel Todd and Hervé Bras, *L'Invention de la France* (Paris: Livre de poche, 1981).

29. *Lettres patentes du roi René de Provence*, Nice, Alpes-Maritimes Departmental Archives, H 0028, 1453.

30. It was not until 1608 that this case was finally closed. See *Requêtes et comparants de l'économie du monastère*, Nice, Alpes-Maritimes Departmental Archives, H 0030, 1581, 1583. Twenty-five setiers equals 3,800 liters, the equivalent of 10,000 pounds or about five metric tons of salt.

31. The Salins saltworks in Franche-Comté enjoyed an uncontested monopoly from the start of the fourteenth century and supplied the neighboring regions of Burgundy, Germany, and Switzerland. Under Louis XVI, 500,000 of 750,000 quintals (one quintal is equivalent to one hundred kilograms) of salt produced annually in eastern France at the time were thus exported. See André Hammerer, *Sur les chemins du sel: Activité commerciale des sauneries de Salins du XIVe au XVIIIe siècle* (Besançon: Cêtre, 1984); Marcel Marion, *Dictionnaire des institutions de la France aux XVIIe–XVIIIe siècles* (Paris: Auguste Picard, 1923), 247–50. Before this, "salt roads" (still called *voies salneresces*) linked the Comtois saltworks (Chaux, Salins, Grozon, Montmorot, Poligny) to Autun, Vergy, and perhaps Dijon. See Jean Richard, "Que sait-on du réseau routier de la Bourgogne au Moyen Age?" *Les Transports et voies de communication: Cahiers de l'Association interuniversitaire de l'Est* 18 (1978): 64. In Lorraine, salt springs or salt wells were found in Rozière, Dieuze, and Château-Salins.

32. Under Louis XVI, a pound of salt sold for twelve to thirteen sous in the provinces of the great gabelle, six to eight sous in the provinces of the

lesser gabelle, two to six sous in the Pays de Salines, less than two sous in the "redeemed provinces," and even less still in the exempt regions. See Marion, *Dictionnaire*, 247–50.

The distinction for tax purposes between the provinces of the greater and lesser gabelles prefigures to a degree the division of France into two zones—free and occupied—under the German occupation during World War II. The price of gasoline represents another case in point: in our day, for a long time it remained significantly lower in the small neighboring countries (Belgium, Luxembourg, Switzerland) than it was in France.

For *portacols*, H. Affre, *Dictionnaire des institutions, mœurs et coutumes du Rouergue* (1903; reprint, Marseilles: Lafitte Reprints, 1974), 407–8.

33. See Marion, *Dictionnaire*.

34. The patrolling boat from the Villefranche tax office was charged with collecting a 2 percent tax for the house of Savoy on vessels crossing the "Sea of Nice" (between the border of the states of Genoa and the mouth of the Var River) and on those putting in at Villefranche. In September 1734 the commander of this ship, his first mate, and a part of the crew were tried for having onloaded 18,000 rups (or 140.6 metric tons) of salt in Sardinia. See Jean-Loup Fontana, "Gabelous et faux saunage sur les côtes de Provence," *L'Entrelus* 3 (1976): 39–44. "A border patrol agent has been arrested in San Diego County, California with more than 500 pounds of marijuana in his car" (dispatch from the *Los Angeles Times* agency, August 8, 1997: Plus ça change . . .

35. Alain Colas, *Le sel*, Que sais-je? no. 339, 2d ed. (Paris: PUF, 1993), p. 18.

36. On this subject, see Emmanuel Le Roy Ladurie and Michel Morineau, *Histoire économique et sociale de la France*, vol. 2, *Paysannerie et croissance*, (Paris: PUF, 1993), 1:746–57.

37. The tradition of the gabelle is continued in indirect taxes, such as those on gas, diesel fuel, fuel oil, cigarettes, and alcohol, and in municipal taxes, such as those on water (so handy for financing political parties in the not too distant past). Once again, plus ça change . . .

38. Joseph Piégay, *Les mulets du sel: Une rébellion paysanne dans le pays du Verdon en 1710* (Mane, Haute-Provence: Alpes de Lumière, 1998).

39. Following a law proposed by René Pléven, finance minister and author of the bill; see Jean Chazelas, "La suppression de la gabelle du sel en 1945," in *Le rôle du sel*, ed. Mollat, 263–65.

40. Sébastien Vauban, *Projet d'une dixme royale* (Brussels: Georges de Backer, 1709), 83.

41. Marion, *Dictionnaire*, 247–50.

42. The latter, perceived as painless, is an irresistibly easy option for governments to take; in OECD nations, "the Value Added Tax has risen from 5

percent of GDP in 1960 to an average of 9 percent in 1988," writes the author of "Le consentement à l'impôt," in ENA, *Sujets et meilleures copies des concours 1994, 1995 et 1996* (Strasbourg: Ecole nationale d'administration, 1997), 103–8. He also writes that "the only solution cannot consist in substituting indirect taxes for direct taxes to increase social acceptability." Already under the ancien régime a kind of VAT had been proposed by Chevalier de Jaucourt; see the article "L'Impôt," in Diderot and d'Alembert's *L'Encyclopédie* (Paris: Editions sociales, 1976), 138–39.

43. This amounts to approximately half of what was planned: the buildings are arranged in a semicircle; Ledoux's design was for the completed circle.

44. The Centre international de réflexion sur le futur (International Center for Reflection on the Future), under the authority of the "Claude-Nicolas Ledoux" Foundation, which was created in 1972 (under a 1901 French law on nonprofit organizations).

45. *Berniers* is a term for saltworkers, elsewhere called *sauniers*. The word *bernier* is a cognate of the English word *brine*, among others.

46. The brine undergoes evaporation, using a method of successive boilings; see Colas, *Le sel*. The successive-boiler building, no longer extant (the saltworks halted its operations at the end of the nineteenth century), one kilometer long, was located between the Loue River, which supplied its hydraulic energy, and the surviving buildings; for more information, see Gérard Marie de Ficquelmont, Olivier Blin, and Claudine Fontanon, eds., *Le Guide du patrimoine industriel scientifique et technique* (Paris: La Manufacture, 1990), 148.

47. T. S. O. Maille, *Sean-Fhocla Chonnacht*, vol. 1 (Dublin, 1948).

48. Gandhi made Sabarmurti his headquarters after his return from South Africa in 1915. This ashram is in Ahmedabad, 600 kilometers southwest of New Delhi; see John F. Burns, "Fifty Years of Freedom: India, the Past to Present. This Birthday Without Party Can Be a Liberation for Visitors," *International Herald Tribune*, August 15, 10.

5. BIOLOGY

1. Mythologies have made efforts at providing answers. Thus a magic millstone continues to turn after having filled a merchant's boat with salt and so salts the sea. See J. Baffie, "Pourquoi l'eau de mer est salée," in *Le sel de la vie en Asie du Sud-Est*, ed. P. Le Roux and J. Ivanoff (Patani: Prince de Songkla University-CNRS, 1993), 365–80.

2. See, for example, Resianne Fontaine, "Why Is the Sea Salty? The Discussion of Salinity in Hebrew Texts of the Thirteenth Century," *Arabic Science and Philosophy* 5 (1995): 195–218.

3. Among other places in *Meteorology*, book 2, part 3.

4. Robert J. P. Williams and J. J. R. Frausto da Silva, *The Natural Selection of the Chemical Elements: The Environment and Life's Chemistry* (New York: Oxford University Press, 1996), 301.

5. Alain Colas, *Le sel*, Que sais-je? no. 339, 2d ed. (Paris: PUF, 1993), 51; Williams and da Silva, *Natural Selection*, 306. Sodium concentrations are three times greater in the Dead Sea and six times greater in the Great Salt Lake. Chlorine concentrations are nine times greater in the Dead Sea and six times greater in the Great Salt Lake.

6. Williams and da Silva, *Natural Selection*, 306.

7. Let's not forget that the ocean is also enriched by volcanic eruptions and by hydrothermal waters springing from between the plates, at the sites of the ocean ridges.

8. Since it is a case of distillation.

9. Williams and da Silva, *Natural Selection*, 300, 311–12.

10. Scott R. Smedley and Thomas Eisner, *Science*, December 15 1995.

11. On the subject as a whole (the need for salt in mammals and humans), see Derek Denton, *The Hunger for Salt* (Berlin: Springer, 1982).

12. To illustrate this, let me just mention a charming Swedish custom: On the Midsummer Night, young unmarried girls would eat salted foods— one or two salted herrings, for example—so as to be thirsty in their sleep that night. Each hoped to see in her dreams a young man who would offer her something to drink. This man was destined to become her future husband.

13. Alain Bombard, *Naufragé volontaire* (Paris: Phébus, D'ailleurs, 1995; Libretto, 1998); *The Voyage of the Hérétique: The Story of Dr. Bombard's 65-day Solitary Atlantic Crossing in a Collapsible Life Raft*, trans. Brian Connell (New York: Simon and Schuster, 1954).

14. Williams and da Silva, *Natural Selection*, 207.

15. Discoveries made by Alan Hodgkin that won him the Nobel Prize in physiology in 1963. See Alan Hodgkin, *Chance and Design* (Cambridge: Cambridge University Press, 1992), esp. ch. 28.

16. Carl Woese and G. E. Fox made this discovery in the 1970s, in examining mitochondrial RNA. See J. Ribier, "Les archéobactéries," in *Le livre de l'année*, ed. F. Trémolières (Paris: Larousse, 1993), 243–44.

17. C. R. Curds, S. S. Banforth, and B. J. Finlay, *Insect Science and Applications* 7 (1986): 447–49.

18. D. Roberts, "Eukaryotes in Extreme Environments" (London: Department of Zoology, Natural History Museum, London, 1996). Their proliferation reddened the water in the brines, in the salt marshes in particular. See, for example, the report by Michaëla Bobasch, "Flâneries en France: A la porte des usines, le touriste curieux," *Le Monde*, August 21, 1997, 14. (One can visit the saltworks in the south of

France near Marseilles by consulting the tourism bureaus in Aigues-Mortes or in Grau-du-Roi.)

19. When the ionic force of the medium increases in greater salt concentrations, the proteins lose their structural integrity and aggregate or precipitate out of the aqueous solution. This occurs because of an increase in hydrophobic forces and because the negative charges of the anionic groupings (acids) are neutralized by the abundance of positive counterions.

20. M. Avron and A. Ben-Amotz, *Strategies of Microbial Life in Extreme Environments*, ed. M. Shilo (Berlin: Dahlem Konferenzen, 1979), 83–91.

21. Williams and da Silva, *Natural Selection*, 386.

22. Jules Henry, *Voyage au pays des Mormons*, ed. E. Dentu, (Paris, 1860), 1:156–158. Translation by the author.

6. OTHER SCIENCE INSIGHTS

1. Laurent d'Houry, *Dictionnaire hermétique* (1695), 179–80.

2. See, for example, Dexter Johnson, "Caustic Soda Prices Surge Ahead Propelled by Shrinking Supplies," *Chemical Market Reporter*, September 1, 1997, 4, 15.

3. Clifford A. Hampel and Gessner G. Hawley, eds., *The Encyclopedia of Chemistry*, 3d ed. (New York: Van Nostrand Reinhold, 1973), s.v. "Sodium."

4. Francis Bacon, *Novum Organum*, ed. T. Fowler (Oxford, 1889), 3, 5.

5. Francis Bacon, *De Augmentis Scientarum*, ed. J. Spedding, vol. 5 of *The Works of Francis Bacon* (Boston: Brown and Taggard, 1857–1859), 2.

6. See Elizabeth David, *Harvest of the Cold Months: The Social History of Ice and Ices* (London: Michael Joseph, 1994), 66.

7. This is one of the uses of salt in the kitchen: to speed up the cooking process by increasing the temperature of boiling water; the other benefit of salted water is to protect the cells—whether animal or vegetal—of a food, which otherwise would risk exploding from water entering through the permeable membranes and diluting the intracellular environment, which consists of a saline aqueous solution.

8. Elizabeth David, "Cathay to Caledonia," in *Harvest of the Cold Months*, 217–45.

9. Félix Singer and W. L. German, *Les Glaçures céramiques* (Saint-Germain-en-Laye: Borax Français, 1960), 38–41.

10. Reported in H. M. Smith, *Torchbearers of Chemistry* (New York: Academic, 1949), 135.

11. Not long after (1869) Lockyer founded the general scientific journal *Nature*, which today competes with *Science* for the announcement of

discoveries great and small, real or fictional (the memory of water). See John Maurice, *La Recherche* (1997): 118–23.

12. R. M. Bonnet, *Les Horizons chimériques* (Paris: Dunod, 1992), 92.

13. Reading an autobiography such as that of Bernard Lovell (*Astronomer by Chance* [Oxford: Oxford University Press, 1992]) demonstrates the determining contribution of wars to the flourishing of knowledge, through the technological developments that they accelerate or make possible.

14. The points at which the sodium channels are grouped along the axon are known as the Ranvier's nodes. The two ends of the chain presented here are held by Andrew F. Huxley and R. Stampfli, "Evidence for Saltatory Conduction in Peripheral Myelinated Fibres," *Journal of Physiology* 108 (1949): 315–36; and by M. R. Kaplan, A. Meyer-Franke, S. Lambert, V. Bennett, I. D. Duncan, S. R. Levinson, and B. A. Barres, "Induction of Sodium Channel Clustering by Oligodendrocytes," *Nature*, April 17, 1997, 724–28.

15. In other languages such as English or Swedish, one says, "to take something with a grain of salt," or "en nypa salt," which denotes a certain skepticism.

16. A good example of this is the notion of the macromolecule. Hermann Staudinger (Nobel Prize in chemistry in 1953) had a great deal of difficulty in introducing it in the 1930s. That a molecule might exceed the dimensions of the unit cell—that is, of the module that is indefinitely repeated in the three dimensions—that contains it seemed altogether absurd. See his autobiography, Hermann Staudinger, *From Organic Chemistry to Macromolecules: A Scientific Autobiography Based on My Original Papers*, trans. Jerome Fock and Michael Fried (New York: Wiley-Interscience, 1970).

17. For the *saugrenu* aspect of the quantification of space that he had postulated, see John S. Rigden, *Rabi, Scientist, and Citizen*, Alfred P. Sloan Foundation Series (New York: Basic, 1987), 48.

18. Or better yet, as the Languedoc saying goes: "Y courre cóumo los fédos o lo sal" (to rush in a crowd like sheep to salt). See Aimé Vayssier, *Dictionnaire patois-français du département de l'Aveyron* (1879; reprint, Marseilles: Lafitte Reprints, 1979), s.v. "Sal."

7. MYTHS

1. For this reason, the Christian host is always made of unleavened bread.

2. The iconography reflects this. Thus, in *The Last Supper*, Leonardo da Vinci places an overturned saltcellar on the table before Judas to symbolize the breaking of the Covenant. See G. Dunoyer de Segonzac, *Les chemins du sel* (Paris: Gallimard, Découvertes, 1991), 111.

3. "Salt," *Encyclopedia Judaica* (Jerusalem, n.d.), 710–11.

4. Since trade in salt was prohibited on Sundays, baptisms consequently could not take place on that day throughout Christendom. See Giovanni Boccaccio, *The Decameron*, trans. John Payne (New York: Horace Liveright, 1925), pt. 2, 153.

5. "You are the salt of the earth; but if salt has lost its taste, how shall its saltness be restored? It is no longer good for anything except to be thrown out and trodden under foot by men" (Matthew 5:13).

6. "Salt," 710–11.

7. Also spelled Hallstatt. This Austrian village on the shores of the Hallstatter See and near the saltworks in the heights of Salzberg (a site listed in the Unesco World Heritage List) is found in the Salzkammergut, the former private lands of the Habsburgs. See Patrick Francès, "Le sel de l'Autriche," *Le Monde*, April 23, 1998, 22.

8. Halmos, another Halia, Halirrhotios, Salacia, and Halesius. See Pierre Grimal, *Dictionnaire de la mythologie grecque et romaine* (Paris: PUF, 1969), esp. 82, 172b, 275a, 318b, 414a.

9. Hesiod, *Theogony*, trans. Hugh G. Evelyn-White, The Perseus Project, The Internet Classics Archive, by Daniel C. Stevenson, Web Atomics, http://classics.mit.edu, verse 245.

10. Georges Dumézil, "Mamurius Veturius," *Tarpeia: Essais de philologie comparative indo-européenne*, La Montagne Sainte-Geneviève (Paris: Gallimard, 1947), 205–46.

11. "Nunc est bibendum ... / ... nunc Saliaribus / ornare pulvinar deorum / tempus est dapibus sodales" (Horace, *Odes*, book 1, ode 37, lines 1–4).

12. "Jam dederat Saliis (a saltu nomina ducant)," *Fastes*, book 3, lines 385–92; "The Greek *als* was a lump of salt, also the sea; *alsis* was something leaping. . . . Latin has . . . also *saliens*, sacrificial salt which, as a good omen, leapt up when thrown into the fire" (Lewis Thomas, *Etcetera Etcetera: Notes of a Word-Watcher* [Harmondsworth, Middlesex: Penguin, 1990], 51).

13. Jacques Soustelle, *The Four Suns: Recollections and Reflections of an Ethnologist in Mexico*, trans. E. Ross (London: Andre Deutsch Limited, 1971), 174.

14. "The triple saltcellar was comprised of three joined salt dishes for coarse salt, fine salt, and pepper; the saltcellar with candleholders could store candlesticks, the saltcellar for eggs included an eggcup" (Nicole Blondel and Catherine Arminjon, *Principes d'analyse scientifique. Objets civils domestiques. Vocabulaire* [Paris: Imprimerie nationale, 1984], Inventaire général des monuments et des richesses artistiques de la France).

15. "Above the salt, below the salt." For example, Ben Jonson wrote: "His fashion is not to take knowledge of him that is beneath him in clothes.

He never drinks below the salt." This means "he doesn't deign to drink with such inferior folk." *Cynthia's Revels*, act 2, scene 2.

16. *The Autobiography of Benvenuto Cellini*, trans. George Bull (London: Folio Society, 1966), 213.

17. Is this the concrete manifestation in image form of the age-old alliteration between the Latin words *sol* and *sal*, which respectively mean "sun" and "salt"?

18. "The saltcellars are covered and even locked until the sixteenth century; they are sometimes on rollers so as to be able to be sent to the various dining guests" (E. Viollet-le-Duc, *Dictionnaire raisonné du mobilier français de l'époque carolingienne à la Renaissance* [Paris: Bance, 1858–1875], 2:149).

19. "Art," in Diderot and D'Alembert, *Encyclopedia: Selections*, trans. Nelly S. Hoyt and Thomas Cassirer (Indianapolis: Bobbs-Merrill, 1965), 5–6.

20. Norah Gillow, *William Morris Designs and Motifs* (London: Studio Editions, 1995), 64.

21. Françoise Jollant, "Le design," in *Grand Larousse annuel. Le livre de l'année*, ed. Patrice Maubourguet (Paris: Larousse, 1994), 239–41.

22. Made by stoneware potters Jeanne and Norbert Pierlot working in the Ratilly chateau in the Puisaye (Yonne). They had sought to revive the craft of stoneware pottery in Saint-Armand-en-Puisaye and impart to it new forms, a modern style of expression linked both to the research of great foreign potters (Bernard Leach, Hamada Shoji) and to contemporary painting and sculpture.

23. For example, in Aveyron "occupying pride of place in a corner [of the hearth], was the salt chest in the form of a bench on which the old people took pleasure in sitting" (Roger Béteille, *La vie quotidienne en Rouergue avant 1914* [Paris: Hachette, 1973], 79).

24. Stendhal, *On Love*, trans. H. B. V. under the direction of C. K. Scott-Moncrieff (New York: Grosset and Dunlap, 1967), xxiii. All page references cited in the text relate to this work.

25. Marie Henri Beyle, *Vie de Henri Brulard*, ed. H. Martineau (Paris, 1927), 2:129, 153, 156, 166, 171.

26. Page 3 of the first essay of the 1826 preface (French text); see Stendhal, *De L'Amour*, ed. Henri Martineau (Paris: Le Divan, 1927).

27. R.-J. Haüy, *Traité élémentaire de physique* (Paris: Courcier, 1806), 1:58.

28. *Poésies diverses* (Paris, 1649), 2, cited by Jean Rousset in vol. 1 of his *Anthologie de la poésie française* (Paris: Armand Colin–Bibliothèque de Cluny, 1961).

29. Last preface (1842).

30. Chaptal, *Eléments de chimie*, 4th ed. (Paris: Deterville, 1803), 37, 33.

31. Cited by Ferdinand Hoefer, *Histoire de la chimie depuis les temps les plus reculés jusqu'à notre époque* (Paris: Hachette, 1842–1843), 2:235.

32. Johann Kepler, *The Six-Cornered Snowflake*, ed. and trans. Colin Hardie (Oxford: Oxford University Press, Clarendon, 1966).

33. R.-J. Haüy, *Traité élémentaire*, 1:57–58.

34. *Questions Naturelles*, ed. René-Just Haüy (Paris: Garnier), 356.

35. Victor Hugo, *The Laughing Man* (London: T. Nelson and Sons, 1908).

36. Victor Hugo, *Notre Dame de Paris*, ed. William Allan Nielson, Harvard Classics Shelf of Fiction (New York: P. F. Collier and Son, 1917), 176.

37. Cited by Bernard Terramorsi, *Henry James; ou, Le sens des profondeurs: Essai sur les nouvelles fantastiques* (Paris: L'Harmattan, 1996).

38. René Guénon, a connoisseur of esoterica and Asian philosophy, confirms that salt represents individuality, the self; see René Guénon, *La Grande Triade*, 3d ed. (Paris: Gallimard, 1957), p. 107.

39. See Pierre Laszlo, *La leçon de choses* (Paris: Austral, Diversio, 1995).